CW00395187

# Nelson Mathematics

for Cambridge
International A Level

# Pure Mathematics

L. Bostock • S. Chandler • T. Jennings

Oxford and Cambridge
leading education together

**OXFORD**
UNIVERSITY PRESS

Great Clarendon Street, Oxford, OX2 6DP, United Kingdom

Oxford University Press is a department of the University of Oxford.
It furthers the University's objective of excellence in research, scholarship,
and education by publishing worldwide. Oxford is a registered trade mark of
Oxford University Press in the UK and in certain other countries

Text © Sue Chandler, Linda Bostock and Trevor Jennings 2012
Original illustrations © Oxford University Press 2014

The moral rights of the authors have been asserted

First published by Nelson Thornes Ltd in 2012
This edition published by Oxford University Press in 2014

British Library Cataloguing in Publication Data
Data available

978-1-4085-1559-4

10 9 8 7 6 5 4 3 2

Printed in China

**Acknowledgements**

**Cover photograph:** Jim Zuckerman/Alamy
**Illustrations:** Tech-Set Ltd, Gateshead
**Page make-up:** Tech-Set Ltd, Gateshead

Previous exam questions from Cambridge International AS and A Level Mathematics 9709 reproduced by permission
of the University of Cambridge Local Examinations Syndicate:

Page 32 Q1: paper 2 question 1 June 2003, Q2: paper 2 question 1 June 2004, Q3: paper 2 question 3 June 2004,
Q4: paper 2 question 1 June 2007, Q5: paper 2 question 2 June 2007, Q6: paper 2 question 1 June 2008, Q7: paper 2
question 2 June 2008, Q8: paper 2 question 4 June 2008, Q9: paper 2 question 1 June 2009, Q10: paper 2 question
2 June 2009, Q11: paper 2 question 6 June 2009; Page 33 Q12: paper 2 question 3 November 2002, Q13: paper 22
question 1 June 2010, Q14: paper 22 question 3 June 2010, Q15: paper 2 question 3 November 2007, Q16: paper 2
question 8 parts i and ii November 2007, Q17: paper 2 question 3 November 2008, Q18: paper 2 question 6 November
2008, Q19: paper 2 question 6 November 2009, Q20: paper 21 question 2 November 2009, Q21: paper 2 question
2 November 2003; Page 71 Q1: paper 2 question 2 June 2004, Q2: paper 2 question 4 June 2004; Page 72 Q3: paper
2 question 3 June 2007, Q4: paper 2 question 7 June 2008, Q5: paper 2 question 7 June 2007, Q6: paper 2 question
5 June 2008, Q7: paper 2 question 5 June 2009, Q8: paper 21 question 2 June 2010, Q9: paper 2 question 3 June
2000, Q10: paper 21 question 7 June 2010; Page 73 Q11: paper 2 question 5 November 2002, Q12: paper 2 question
7 November 2002, Q13: paper 2 question 7 November 2003, Q14: paper 2 question 8 November 2007, Q15: paper 2
question 7 November 2007, Q16: paper 2 question 8 November 2009, Q17: paper 21 question 4 November 2008, Q18:
paper 2 question 6 November 2009; Page 131 Q1: paper 3 question 5 June 2003, Q2: paper 3 question 6 June 2003, Q3:
paper 3 question 7 June 2003; Page 132 Q4: paper 3 question 9 June 2003, Q5: paper 3 question 6 June 2004, Q6: paper
3 question 8 June 2004, Q7: paper 3 question 9 June 2004, Q8: paper 3 question 10 June 2004, Q9: paper 3 question
1 June 2007, Q10: paper 3 question 3 June 2007, Q11: paper 3 question 4 June 2007, Q12: paper 3 question 5 June
2007, Q13: paper 3 question 10 June 2007; Page 133 Q14: paper 3 question 10 June 2008, Q15: paper 3 question 3 June
2008, Q16: paper 3 question 4 June 2008, Q17: paper 3 question 5 June 2009, Q18: paper 3 question 2 June 2009, Q19:
paper 3 question 7 June 2009, Q20: paper 3 question 8 June 2009; Page 134 Q21: paper 3 question 9 June 2009, Q22:
paper 32 question 6 June 2010, Q23: paper 32 question 8 June 2010, Q24: paper 32 question 9 June 2010, Q25: paper
32 question 10 June 2010, Q26: paper 31 question 10 November 2009, Q27: paper 31 question 4 November 2009,
Q28: paper 31 question 3 November 2009; Page 135 Q29: paper 31 question 8 November 2009, Q30: paper 3 question
9 November 2008, Q31: paper 3 question 7 November 2008, Q32: paper 3 question 2 November 2008, Q33: paper 3
question 5 November 2008

The University of Cambridge Local Examinations Syndicate bears no responsibility for the example answers to
questions taken from its past question papers which are contained in this publication.

# P2&3 Contents

# Introduction

The *Nelson Mathematics for Cambridge International A Level* series has been written specifically for students of Cambridge's 9709 syllabus by an experienced author team in collaboration with examiners who are very familiar with the syllabus and examinations. This means that, no matter which combination of modules you have chosen, the content of this series matches the content of the syllabus exactly and will give you firm guidelines on which to base your studies.

This book is designed for candidates studying either the Pure Mathematics 2 or Pure Mathematics 3 module. It is divided into 15 chapters that give a sensible order for your studies. The chapters begin with a list of objectives that show you what is covered. Chapters 1–10 cover content required for both P2 and P3 modules, Chapters 11–15 cover content that only applies to the P3 module. Throughout the exercises, questions have been marked with a P3 symbol if they are particularly suitable for P3 candidates.

The following features help you to understand the concepts of the P2 and P3 modules and to succeed in your exams.

- The introductions to concepts are accompanied by examples of questions together with their solutions. These show each step of working along with a commentary on the reasoning processes involved.

- There are numerous exercises for you to practise what you have learned and develop your skills.

- There are three Summary exercise sections with more detailed questions covering the content of the preceding chapters. These questions are similar to those found in exam papers and all are from real exam papers.

- Summaries of key information and formulae are at convenient points in the book to help you revise what you have covered in the last few chapters.

- Answers to all questions are provided at the back of the book for you to check your answers to exercises.

- Two sample examination papers have been created in the style of Cambridge's International A Level P2 and P3 exams to give you the experience of working through a full examination paper.

# 1 Algebra

*After studying this chapter you should be able to*

- divide a polynomial, of degree not exceeding 4, by a linear or quadratic polynomial, and identify the quotient and remainder (which may be zero)
- use the factor theorem and the remainder theorem, e.g. to find factors, solve polynomial equations or evaluate unknown coefficients.

## DIVIDING A POLYNOMIAL BY A LINEAR OR QUADRATIC EXPRESSION

A polynomial is a function of the form $f(x) = a_1 x^n + a_2 x^{n-1} + a_3 x^{n-2} + \ldots a_{n-1} x + a_n$, $x \in \mathbb{R}$ where $n$ is an integer and $a_1, a_2, \ldots$ are rational constants, $a_1 \neq 0$.

A fraction where both the numerator and the denominator are polynomials, is *proper* if the highest power of $x$ in the numerator is less than the highest power of $x$ in the denominator, e.g. $\dfrac{x+1}{x^2 + x + 2}$.

When this is not the case the fraction is *improper* and we can divide the numerator by the denominator using long division.

The following example shows how you divide $x + 2$ into $x^2 + 3x - 7$.

($x^2 + 3x - 7$ is called the *dividend* and $x + 2$ is called the *divisor*.)

$$
\begin{array}{r}
x + 1 \\
x + 2 \overline{)\, x^2 + 3x - 7} \\
\underline{x^2 + 2x} \\
x - 7 \\
\underline{x + 2} \\
-9
\end{array}
$$

Start by dividing $x$ into $x^2$; it goes $x$ times.
Multiply $x + 2$ by $x$ and then subtract this from $x^2 + 3x$.
Bring down the $-7$, divide $x$ into $x$ and repeat the process.
No more division by $x$ can be done, so stop here.

$x + 1$ is called the *quotient* and the *remainder* is $-9$.

This compares with dividing, say, 32 by 5, where the quotient is 6 and the remainder is 2.

But writing $\frac{32}{5}$ as a mixed number gives $6 + \frac{2}{5}$.

Similarly $\dfrac{x^2 + 3x - 7}{x + 2} \equiv x + 1 - \dfrac{9}{x + 2}$

where $x + 1$ is a linear function and $\dfrac{9}{x + 2}$ is a rational fraction.

---

### Examples 1a

**1** Divide $2x^3 - x + 5$ by $x + 3$

$$
\begin{array}{r}
2x^2 - 6x + 17 \\
x + 3 \overline{)\, 2x^3 - 0x^2 - x + 5} \\
\underline{2x^3 + 6x^2} \\
-6x^2 - x \\
\underline{-6x^2 - 18x} \\
17x + 5 \\
\underline{17x + 51} \\
-46
\end{array}
$$

$x$ into $2x^3$ goes $2x^2$ times.
Multiply $2x^2$ by $x + 3$ and subtract.
Bring down $-x$.
$x$ into $-6x^2$ goes $-6x$ times.
Bring down 5.
$x$ into $17x$ goes 17 times.

The quotient is $2x^2 - 6x + 17$ and the remainder is $-46$.

When the divisor is quadratic, the same method can be used.

When there are missing terms, e.g. no term in $x$, the division is easier if they are included with a coefficient of zero.

---

### Examples 1a cont.

2  Divide $x^3 + 4x^2 - 7$ by $x^2 - 3$

There is no $x$ term in either polynomial so add the term $0x$ to each polynomial.

$$
\begin{array}{r}
x + 4 \quad\longleftarrow\ \text{the quotient} \\
x^2 + 0x - 3\ \overline{\smash{)}\ x^3 + 4x^2 + 0x - 7} \\
\underline{x^3 \phantom{+ 4x^2} - 3x} \\
4x^2 + 3x - 7 \\
\underline{4x^2 \phantom{+ 3x} - 12} \\
3x + 5 \quad\longleftarrow\ \text{the remainder}
\end{array}
$$

$x^2$ into $x^3$ goes $x$ times.
Multiply $x^2 + 0x - 3$ by $x$ and subtract.
$x^2$ into $4x^2$ goes 4 times.
Multiply $x^2 + 0x - 3$ by 4 and subtract.

giving $\dfrac{x^3 + 4x^2 - 7}{x^2 - 3} = x + 4 + \dfrac{3x + 5}{x^2 - 3}$

---

### Exercise 1a

Carry out each of the following divisions, giving the quotient and the remainder.

1  $(2x^2 + 5x - 3) \div (x + 2)$

2  $(x^2 - x + 4) \div (x + 1)$

3  $(4x^3 + x - 1) \div (2x - 1)$

4  $(2x^3 - x^2 + 2) \div (x - 2)$

5  $(x^3 + x^2 - 3x + 6) \div (x^2 + 3)$

6  $(x^4 + 5x^2 + 2) \div (x + 1)$

7  $(3x^3 - 5) \div (x - 2)$

8  $(3x^4 - 2x^2 + 10) \div (x^2 - x)$

9  $(x^4 + x^3 - 2x^2 + 5x - 1) \div (x^2 + 4)$

10  $(5x^4 + x^3 - 1) \div (x^2 + 3x + 4)$

## THE REMAINDER THEOREM

When $f(x) = x^3 - 7x^2 + 6x - 2$ is divided by $x - 2$, we get a quotient and a remainder. The relationship between these quantities can be written as

$f(x) = x^3 - 7x^2 + 6x - 2 \equiv (\text{quotient})(x - 2) + \text{remainder}$

Substituting 2 for $x$ eliminates the term containing the quotient, giving

$f(2) = \text{remainder}$

This is an illustration of the general case: when a polynomial $f(x)$ is divided by $(x - a)$ then

$f(x) \equiv (\text{quotient})(x - a) + \text{remainder}$

$\Rightarrow \quad f(a) = \text{remainder}$

This result is called the *remainder theorem* and can be summarised as

**when a polynomial f(x) is divided by (x − a), the remainder is f(a).**

### Examples 1b

1  Find the remainder when

(a)  $x^3 - 2x^2 + 6$ is divided by $x + 3$

(b)  $6x^2 - 7x + 2$ is divided by $2x - 1$

(a)  When $f(x) = x^3 - 2x^2 + 6$ is divided by $x + 3$, the remainder is
$$f(-3) = (-3)^3 - 2(-3)^2 + 6 = -39$$

(b)  If  $f(x) = 6x^2 - 7x + 2$, then
$$f(x) = (2x - 1)(\text{quotient}) + \text{remainder}$$
$$\Rightarrow \quad \text{remainder} = f\left(\tfrac{1}{2}\right) = 0 \qquad \text{Note that as the remainder is zero, } 2x - 1 \text{ is a factor of } f(x).$$

## THE FACTOR THEOREM

When $(x - a)$ is a factor of the polynomial $f(x)$, there is no remainder $\Rightarrow$ $f(a) = 0$

This is the factor theorem.

**i.e.   if, for a polynomial f(x), f(a) = 0 then x − a is a factor of f(x).**

For example, when $x = 3$,
$$x^4 - 3x^3 - 3x^2 + 11x - 6 = 81 - 81 - 27 + 33 - 6 = 0$$

Therefore $x - 3$ is a factor of $x^4 - 3x^3 - 3x^2 + 11x - 6$.

The factor theorem can be used to
- find factors of polynomials
- find unknown constants in a polynomial when a factor of that polynomial is known
- solve equations of the form $f(x) = 0$ where $f(x)$ is a polynomial.

### Examples 1b cont.

2  Find the value of $a$ for which $2x - 1$ is a factor of $4x^3 - 2x^2 + ax - 4$

As $2x - 1$ is a factor of $4x^3 - 2x^2 + ax - 4$,
$$4\left(\tfrac{1}{2}\right)^3 - 2\left(\tfrac{1}{2}\right)^2 + a\left(\tfrac{1}{2}\right) - 4 = 0$$

i.e.  $\tfrac{1}{2} - \tfrac{1}{2} + \tfrac{1}{2}a - 4 = 0 \qquad \therefore \quad a = 8$

Using the factor theorem we know that the value of the expression is zero when $x = \tfrac{1}{2}$ (the value of $x$ for which $2x - 1 = 0$).

3  Factorise $x^4 - 16$

$$\begin{aligned} x^4 - 16 &= (x^2 - 4)(x^2 + 4) \\ &= (x - 2)(x + 2)(x^2 + 4) \end{aligned}$$

Using the factor theorem will find factors if they exist, but it is not always the quickest method. Look for forms that can be recognised. In this case $x^4 - 16$ is the difference of two squares.

**4** The equation $f(x) = 0$ has a repeated root, where $f(x) = 4x^2 + px + q$. When $f(x)$ is divided by $x + 1$ the remainder is 1. Find the values of $p$ and $q$

$$f(-1) = 4 - p + q = 1 \quad \Rightarrow \quad p = q + 3 \qquad\qquad\qquad [1]$$

If $4x^2 + px + q = 0$ has a repeated root then '$b^2 - 4ac$' $= 0$

i.e. $p^2 - 16q = 0$ $\qquad\qquad\qquad\qquad\qquad\qquad\qquad\qquad\qquad [2]$

Solving equations [1] and [2] simultaneously gives

$$(q + 3)^2 - 16q = 0 \quad \Rightarrow \quad q^2 - 10q + 9 = 0 \quad \Rightarrow \quad (q - 9)(q - 1) = 0$$

$\therefore\quad$ either $q = 9$ and $p = 12$ or $q = 1$ and $p = 4$

---

**5** Factorise $2x^3 - x^2 - 2x + 1$. Hence solve the equation $2x^3 - x^2 - 2x + 1 = 0$

If $ax + b$ is a factor of $2x^3 - x^2 - 2x + 1$, then $(ax + b)(cx^2 + dx + e) = 2x^3 - x^2 - 2x + 1$
The term $2x^3$ comes from $ax \times cx^2$ so possible values of $a$ are 1 (in which case $c = 2$) or 2 (in which case $c = 1$).
The term $+1$ comes from $b \times e$ so possible values of $b$ are $\pm 1$ (in which case $e = \pm 1$).
Therefore possible factors are $x \pm 1$ and $2x \pm 1$. Try each in turn until one factor is found.

When $x = 1$, $2x^3 - x^2 - 2x + 1 = 2 - 1 - 2 + 1 = 0$

$\therefore\quad (x - 1)$ is a factor

As $x - 1$ is a factor, $2x^3 - x^2 - 2x + 1 = (x - 1)(ax^2 + bx + c)$.
The values of $a$ and $c$ can be written down directly as $(x)(ax^2) = 2x^3$ so $a = 2$ and $(-1)(c) = 1$ so $c = -1$

$$\Rightarrow\quad 2x^3 - x^2 - 2x + 1 = (x - 1)(2x^2 + bx - 1)$$

The term in $x^2$ in the expansion comes from $(-1)(2x^2) + (x)(bx)$

Comparing the coefficients of $x^2$ gives $-1 = -2 + b \quad \Rightarrow \quad b = 1$

$\therefore\quad 2x^3 - x^2 - 2x + 1 = (x - 1)(2x^2 + x - 1)$

The quadratic factor can also be found by dividing $2x^3 - x^2 - 2x + 1$ by $x - 1$.

$$= (x - 1)(2x - 1)(x + 1)$$

$2x^3 - x^2 - 2x + 1 = 0 \quad \Rightarrow \quad (x - 1)(2x - 1)(x + 1) = 0$

$\therefore\quad x = 1 \quad$ or $\quad \frac{1}{2} \quad$ or $\quad -1$

---

**6** Show that $x - 1$ is a factor of $x^4 - x^3 + 2x - 2$

Hence solve the equation $x^4 - x^3 + 2x - 2 = 0$

When $x = 1$, $x^4 - x^3 + 2x - 2 = 1 - 1 + 2 - 2 = 0$

Therefore $x - 1$ is a factor of $x^4 - x^3 + 2x - 2$.

$$x^4 - x^3 + 2x - 2 = (x - 1)(x^3 + bx^2 + cx + 2)$$

To solve the equation $x^4 - x^3 + 2x - 2 = 0$ we need to factorise $x^4 - x^3 + 2x - 2$ completely.

Using the fact that $x^4$ comes from multiplying $x$ by the term in $x^3$ in the cubic factor, so this term must be just $x^3$ and using the fact that $-2$ comes from multiplying $-1$ by the constant term in the cubic factor, so the constant must be $+2$.

Expanding $(x - 1)(x^3 + bx^2 + cx + 2)$ gives

$$x^4 - x^3 + 2x - 2 = x^4 + (b - 1)x^3 + (c - b)x^2 + (2 - c)x + 2$$

Comparing coefficients of $x^3$ gives $-1 = b - 1$ so $b = 0$

Comparing coefficients of $x$ gives $2 = 2 - c$ so $c = 0$

This does not need to be written down, it can be done mentally.

Therefore $x^4 - x^3 + 2x - 2 = (x - 1)(x^3 + 2)$

Possible factors of $f(x) = x^3 + 2$ are $x \pm 1$ and $x \pm 2$.

$f(1) \neq 0, f(-1) \neq 0, f(2) \neq 0, f(-2) \neq 0$

Therefore $x^3 + 2$ has no factors.

Hence $x^4 - x^3 + 2x - 2 = 0$

gives $(x - 1)(x^3 + 2) = 0$

so $x = 1$ is the only solution.

## Exercise 1b

**1** Find the remainder when the following polynomials are divided by the given factors.

(a) $x^3 - 2x + 4, x - 1$

(b) $x^3 + 3x^2 - 6x + 2, x + 2$

(c) $2x^3 - x^2 + 2, x - 3$

(d) $x^4 - 3x^3 + 5x, 2x - 1$

(e) $9x^5 - 5x^2, 3x + 1$

(f) $x^3 - 2x^2 + 6, x - a$

(g) $x^2 + ax + b, x + c$

(h) $x^4 - 2x + 1, ax - 1$

**2** If $x^2 - 7x + a$ has a remainder 1 when divided by $x + 1$, find $a$.

**3** Find whether $x - 1$ is a factor of $x^3 - 7x + 6$

**4** Is $x + 1$ a factor of $x^3 - 2x^2 + 1$?

**5** Show that $2x - 1$ is a factor of $2x^4 - x^3 + 6x^2 - x - 1$

**6** Determine whether $x - 3$ and/or $2x + 1$ are factors of $4x^3 - 7x + 9$

**7** Show that $x - 3$ is a factor of $x^3 - 7x - 6$

**8** Factorise fully

(a) $x^3 + 2x^2 - x - 2$

(b) $x^3 - x^2 - x - 2$

(c) $2x^3 - x^2 + 2x - 1$

(d) $x^4 - 81$

(e) $x^3 + 27$

(f) $x^4 + x^3 - 3x^2 - 4x - 4$

**9** Use your answers to question **8**, parts (a), (d), (e) and (f) to solve the equations

(a) $x^3 + 2x^2 - x - 2 = 0$

(b) $x^4 - 81 = 0$

(c) $x^3 + 27 = 0$

(d) $x^4 + x^3 - 3x^2 - 4x - 4 = 0$

**10** $x - 4$ is a factor of $x^3 - ax + 16$
Find $a$.

**11** $(x + 1)$ and $(x + 2)$ are both factors of $2x^3 + bx^2 - 5x + c$
Find the values of $b$ and $c$.

**12** $x^3 - 4x^2 - 25$ has a factor $(x - a)$

(a) Find the value of $a$.

(b) Solve the equation $x^3 - 4x^2 - 25 = 0$

**13** Solve the equation $x^3 - 3x^2 - 4x + 12 = 0$

**14** Divide $x^3 - 4x^2 + 5$ by $x - 1$, giving the quotient and the remainder.

P3 **15** Given that $(x - 1)$ and $(x + 2)$ are factors of $x^3 + ax^2 + bx - 6$, find the values of $a$ and $b$.

P3 **16** A function f is defined by

$$f(x) = 5x^3 - px^2 + x - q$$

When $f(x)$ is divided by $x - 2$, the remainder is 3. Given that $(x - 1)$ is a factor of $f(x)$

(a) find $p$ and $q$

(b) find the number of real roots of the equation

$$5x^3 - px^2 + x - q = 0$$

P3**17** The function f is given by

$f(x) = 6x^3 - 5x^2 + 5x - a$

$3x^2 - x + 2$ is a factor of f($x$).

(a) Find the value of $a$.

(b) Sketch the curve $y = $ f($x$) showing where the curve cuts the $x-$axis.

P3**18** Show that $x^2 - 3x + 2$ is a factor of

$x^4 - 3x^3 + 3x^2 - 3x + 2$

Hence solve the equation

$x^4 - 3x^3 + 3x^2 - 3x + 2 = 0$

P3**19** The function f is given by

$f(x) = 3x^3 + ax^2 + bx - 1$

$x + 1$ is a factor of f($x$) and of f$'(x)$.

(a) Find the values of $a$ and $b$.

(b) Factorise f($x$) completely.

P3**20** The polynomial $x^4 - 3x^3 - 10x^2 + 27x + a$ is divisible by $x^2 - 3x - 1$

Find the other quadratic factor of the polynomial.

# 2 Logarithms

## After studying this chapter you should be able to

- understand the relationship between logarithms and indices, and use the laws of logarithms
- use logarithms to solve equations of the form $a^x = b$, and similar inequalities.

## INDICES

Logarithms depend on the laws of indices so here is a reminder of these laws.

- $a^p \times a^q = a^{p+q}$              For example, $x^3 \times x^4 = x^{3+4} = x^7$
- $a^p \div a^q = a^{p-q}$              For example, $x^3 \div x^4 = x^{3-4} = x^{-1}$
- $(a^p)^q = a^{pq}$              For example, $(x^3)^4 = x^{3\times4} = x^{12}$
- $a^0 = 1, \ a^{-n} = \dfrac{1}{a^n}, \ a^{\frac{1}{n}} = \sqrt[n]{a}$

## LOGARITHMS

We can read the statement $10^2 = 100$ as

> the base 10 raised to the power 2 gives 100.

This relationship can be rearranged to give the same information, i.e.

2 is the power to which the base 10 must be raised to give 100.

**In this form the power is called a logarithm (log).**

The whole relationship can then be abbreviated to read

2 is the logarithm to the base 10 of 100

or          $2 = \log_{10} 100$

In the same way,          $2^3 = 8$          $\Rightarrow$          $3 = \log_2 8$

and                    $3^4 = 81$          $\Rightarrow$          $4 = \log_3 81$

Similarly          $\log_5 25 = 2$          $\Rightarrow$          $25 = 5^2$

and                    $\log_9 3 = \frac{1}{2}$          $\Rightarrow$          $3 = 9^{\frac{1}{2}}$

Although we have so far used only certain bases, the base of a logarithm can be any positive number, or even an unspecified number represented by a letter, for example

$$b = a^c \qquad \Leftrightarrow \qquad \log_a b = c$$

The symbol $\Leftrightarrow$ means that each of these facts implies the other.

Also as $a^0 = 1$     $\Rightarrow$     $\log_a 1 = 0$,     i.e the logarithm of 1 to any base is zero.

The power of a positive number always gives a positive result

e.g.          $4^2 = 16, 4^{-2} = \frac{1}{16}, \ldots$

This means that, if $\log_a b = c$, i.e. $b = a^c$, then $b$ must be positive, so logs of positive numbers exist, but

**the logarithm of a negative number does not exist.**

## Example 2a

(a)  Write $\log_2 64 = 6$ in index form.

(b)  Write $5^3 = 125$ in logarithmic form.

(c)  Complete the statement $2^{-3} = ?$ and then write it in logarithmic form.

(a)  If $\log_2 64 = 6$ then the base is 2, the number is 64 and the power (i.e. the log) is 6.

$$\log_2 64 = 6 \qquad \Rightarrow \qquad 64 = 2^6$$

(b)  If $5^3 = 125$ then the base is 5, the log (i.e. the power) is 3 and the number is 125.

$$5^3 = 125 \qquad \Rightarrow \qquad 3 = \log_5 125$$

(c)  $2^{-3} = \frac{1}{8}$

The base is 2, the power (log) is $-3$ and the number is $\frac{1}{8}$.

$$2^{-3} = \frac{1}{8} \qquad \Rightarrow \qquad -3 = \log_2 \left( \frac{1}{8} \right)$$

## Exercise 2a

Convert each of the following facts to logarithmic form.

**1**  $10^3 = 1000$

**2**  $2^4 = 16$

**3**  $10^4 = 10\,000$

**4**  $3^2 = 9$

**5**  $4^2 = 16$

**6**  $5^2 = 25$

**7**  $10^{-2} = 0.01$

**8**  $9^{\frac{1}{2}} = 3$

**9**  $5^0 = 1$

**10**  $4^{\frac{1}{2}} = 2$

**11**  $12^0 = 1$

**12**  $8^{\frac{1}{3}} = 2$

**13**  $p = q^2$

**14**  $x^y = 2$

**15**  $p^q = r$

Convert each of the following facts to index form.

**16**  $\log_{10} 100\,000 = 5$

**17**  $\log_4 64 = 3$

**18**  $\log_{10} 10 = 1$

**19**  $\log_2 4 = 2$

**20**  $\log_2 32 = 5$

**21**  $\log_{10} 1000 = 3$

**22**  $\log_5 1 = 0$

**23**  $\log_3 9 = 2$

**24**  $\log_4 16 = 2$

**25**  $\log_3 27 = 3$

**26**  $\log_{36} 6 = \frac{1}{2}$

**27**  $\log_a 1 = 0$

**28**  $\log_x y = z$

**29**  $\log_a 5 = b$

**30**  $\log_p q = r$

## Evaluating logarithms

It is easier to solve a simple equation in index form than in log form so we often use an index equation in order to evaluate a logarithm. For example to evaluate $\log_{49} 7$ we can say

if $\quad x = \log_{49} 7 \quad$ then $\quad 49^x = 7 \quad \Rightarrow \quad x = \frac{1}{2}$

therefore $\qquad \log_{49} 7 = \frac{1}{2}$

In particular, for any base $b$,

if $\quad x = \log_b 1 \quad$ then $\quad b^x = 1 \quad \Rightarrow \quad x = 0$

**i.e. the logarithm of 1 to any base is zero.**

## Using a calculator

A scientific calculator can be used to find the values of logarithms with a base of 10.

The button marked $\boxed{\log}$ gives the value of a logarithm with a base of 10.

You can assume that when the base is not given it can be taken as 10. For example, log 5 means $\log_{10} 5$.

### Exercise 2b

Evaluate

1 $\log_2 4$

2 $\log_{10} 1\,000\,000$

3 $\log_2 64$

4 $\log_3 81$

5 $\log_8 64$

6 $\log_4 64$

7 $\log_9 3$

8 $\log_{\frac{1}{2}} 4$

9 $\log_{10} 0.1$

10 $\log_{121} 11$

11 $\log_5 1$

12 $\log_2 2$

13 $\log_{64} 4$

14 $\log_{99} 1$

15 $\log_{27} 3$

16 $\log_a a^3$

17 $\log_b b^3$

## LAWS OF LOGARITHMS

Given $\qquad x = \log_a b \qquad$ and $\qquad y = \log_a c$

then $\quad a^x = b \qquad$ and $\quad a^y = c$

Now $\qquad\qquad bc = (a^x)(a^y)$

$\Rightarrow \qquad\qquad bc = a^{x+y}$

Therefore $\quad \log_a bc = x + y$

i.e. $\qquad \log_a bc = \log_a b + \log_a c$

This is the first law of logarithms and, as $a$ can represent *any* base, this law applies to the log of *any* product *provided that the same base is used for all the logarithms in the formula.*

Using $x$ and $y$ again, a law for the log of a fraction can be found.

$$\frac{b}{c} = \frac{a^x}{a^y} \qquad \Rightarrow \qquad \frac{b}{c} = a^{x-y}$$

Therefore $\qquad \log_a\left(\frac{b}{c}\right) = x - y$

i.e $\qquad\qquad \log_a\left(\frac{b}{c}\right) = \log_a b - \log_a c$

A third law allows us to deal with an expression of the type $\log_a b^n$

Using $\qquad\qquad x = \log_a b^n \qquad \Rightarrow \qquad a^x = b^n$

i.e. $\qquad\qquad a^{\frac{x}{n}} = b$

Therefore $\qquad\quad \dfrac{x}{n} = \log_a b \qquad \Rightarrow \qquad x = n\log_a b$

i.e. $\qquad\qquad \log_a b^n = n\log_a b$

We now have the three most important laws of logarithms. Because they are true for *any* base we do not need to include a base in the formula but

**in each of these laws every logarithm must be to the same base.**

$$\log bc = \log b + \log c$$

$$\log \frac{b}{c} = \log b - \log c$$

$$\log b^n = n\log b$$

---

### Examples 2c

**1** Express $\log pq^2 \sqrt{r}$ in terms of $\log p$, $\log q$ and $\log r$

$$\log pq^2 \sqrt{r} = \log p + \log q^2 + \log \sqrt{r}$$
$$= \log p + 2\log q + \tfrac{1}{2}\log r$$

**2** Simplify $3\log p + n\log q - 4\log r$

$$3\log p + n\log q - 4\log r = \log p^3 + \log q^n - \log r^4$$
$$= \log \frac{p^3 q^n}{r^4}$$

## Exercise 2c

Express in terms of log $p$, log $q$ and log $r$

1  $\log pq$

2  $\log pqr$

3  $\log \dfrac{p}{q}$

4  $\log \dfrac{pq}{r}$

5  $\log \dfrac{p}{qr}$

6  $\log p^2 q$

7  $\log \dfrac{q}{r^2}$

8  $\log p\sqrt{q}$

9  $\log \dfrac{p^2 q^3}{r}$

10  $\log \sqrt{\dfrac{q}{r}}$

11  $\log q^n$

12  $\log p^n q^m$

Simplify

13  $\log p + \log q$

14  $2 \log p + \log q$

15  $\log q - \log r$

16  $3 \log q + 4 \log p$

17  $n \log p - \log q$

18  $\log p + 2 \log q - 3 \log r$

## EQUATIONS CONTAINING LOGARITHMS OR $x$ AS A POWER

When $x$ forms part of an index, first look to see whether the value of $x$ is obvious.
For example, when $4^x = 16$, then $x = 2$ since $4^2 = 16$

Slightly less obvious is the equation $4^x = 32$

In this case, 4 and 32 can both be expressed as powers of 2;

i.e.     $4^x = 32$     $\Rightarrow$     $(2^2)^x = 2^5$,     i.e. $2^{2x} = 2^5$

so     $2x = 5$,     i.e. $x = 2.5$

When the value of the unknown is not so obvious, taking logs will often transform the index into a factor.

For example, when $5^x = 10$, taking logs of both sides gives

$$x \log 5 = \log 10 \quad \Rightarrow \quad x = \frac{\log 10}{\log 5} = 1.43 \text{ (3 s.f.)}$$     Remember that $\dfrac{\log 10}{\log 5}$ is NOT equal to $\log \dfrac{10}{5}$

When an equation contains logs involving $x$, first look to see whether there is an obvious solution, for example, when $\log_a x = \log_a 4$ it is clear that $x = 4$

When the solution is not so obvious, express the log terms as a single log and then remove the logarithms.

For example, when $2 \log_2 x - \log_2 8 = 1$

using the laws of logs gives $\log_2 \dfrac{x^2}{8} = 1$

then removing the logs gives $\dfrac{x^2}{8} = 2^1$     $\Rightarrow$     $x^2 = 16$     $\Rightarrow$     $x = 4$

($x = -4$ is not a solution because $\log_2 (-4)$ does not exist.)

This shows that it is essential, when solving equations involving logs or indices, that all roots are checked in the original equation.

# INEQUALITIES CONTAINING $x$ AS A POWER

An inequality that contains $x$ as a power can be treated in a similar way to an equation. Remember to look for an obvious solution first. For example if $4^x < 16$, then $x < 2$

Remember that an inequality remains true when you add and subtract any number on both sides, or multiply both sides by a positive number. If you multiply both sides by a negative number, the inequality is reversed.

For example, for $\qquad 3^x < 2$

taking logs gives $\qquad x \log 3 < \log 2$

Therefore $\qquad x < \dfrac{\log 2}{\log 3}$

Log 3 is positive, so dividing both sides by log 3 does not change the inequality. Be careful when dividing by a logarithm, if it is negative it will change the inequality.

An answer to an inequality should be left in exact form, as evaluating it as a decimal and then rounding the answer can give an incorrect result.

For example, $\qquad x < \dfrac{\log 2}{\log 3} \quad \Rightarrow \quad x < 0.630929\ldots$

Rounding 0.630929… to three significant figures gives 0.631. But 0.631 is greater than 0.6309…, so it is not true to say that $x < 0.631$

## Exercise 2d

Solve the equations in questions **1** to **11**.

**1**  $3^x = 9$

**2**  $3^x = \frac{1}{9}$

**3**  $9^x = 27$

**4**  $3^x = 6$

**5**  $2^{2x} = 5$

**6**  $5^x = 4$

**7**  $3^{x-1} = 7$

**8**  $4^{2x+1} = 8$

**9**  $\log_2 x = \log_2 (2x - 1)$

**10**  $\log_4 x = 2$

**11**  $\log x = 2 \log (x - 2)$

Solve the inequalities in questions **12** to **16**.

**12**  $3^x > 27$

**13**  $2^x < 32$

**14**  $5^{x-2} > 125$

**15**  $3^x > 10$

**16**  $2^{2x} < 3 \times 2^{3x}$

**17**  Express $\log_x 5 - 2 \log_x 3$ as a single log term. Hence find the value of $x$ when $\log_x 5 - 2 \log_x 3 = 2$

**18**  Express $\log_3 y - 2 \log_3 x$ as a single logarithm. Hence express $y$ in terms of $x$ when $\log_3 y - 2 \log_3 x = 1$

**19**  Given that $y = 2^x$, express $2^{2x}$ in terms of $y$. By substituting $y$ for $2^x$, solve the equation $2^{2x} - 2^x - 2 = 0$

**P3 20**  Use the substitution $y = a^x$ or otherwise to solve the equation $a^{3x} - 2a^{2x} + a^x = 0$

**P3 21**  Solve the equation $5^x = 5^{x+2} - 5^3$

# 3 Exponential and logarithmic functions

## After studying this chapter you should be able to

- understand the definition and properties of $e^x$ and $\ln x$, including their relationship as inverse functions and their graphs
- use the derivatives of $e^x$, $\ln x$ together with constant multiples, sums, differences and composites
- use logarithms to transform a given relationship to linear form, and hence determine unknown constants by considering the gradient and/or intercept.

## EXPONENTIAL CURVES

The curve $y = a^x$ for $a > 0$ is called an exponential curve. The diagram shows the graphs of some members of the exponential family of curves.

These curves all go through the point $(0, 1)$.
All curves in the family go through this point
because, when $y = a^x$ for any positive value of $a$,

$$x = 0 \quad \Rightarrow \quad y = a^0 = 1$$

Each of these curves has a property that can
be found from an accurate plot of the graph by
drawing the tangents at a few points on the graph.
When the gradients of these tangents are

calculated, the value of $\dfrac{dy}{dx} \div y$ is constant.

The table gives approximate values of this
constant for different values of $a$.

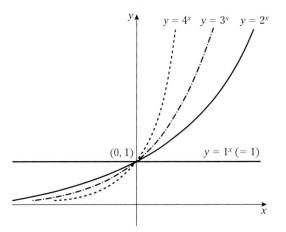

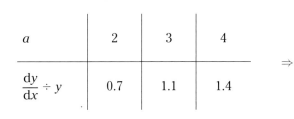

| $a$ | 2 | 3 | 4 |
|---|---|---|---|
| $\dfrac{dy}{dx} \div y$ | 0.7 | 1.1 | 1.4 |

$\Rightarrow$

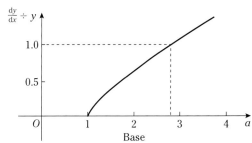

The graph shows values of $\dfrac{dy}{dx} \div y$ against values of $a$.

The graph shows that somewhere between $a = 2$ and $a = 3$, there is a number for which

$$\frac{dy}{dx} \div y = 1, \quad \text{i.e.} \quad y = \frac{dy}{dx}$$

This number is called e.
e is irrational; i.e. like $\pi$, $\sqrt{2}$, etc., it cannot be given an exact decimal value but, to 4 significant figures,
e = 2.718

Therefore when $y = e^x$, $\dfrac{dy}{dx} = e^x$

# EXPONENTIAL FUNCTIONS

$a^x$ has a single value for every real value of $x$. Therefore $x \mapsto a^x$ is a function for $x \in \mathbb{R}$

  $f(x) = a^x$, $x \in \mathbb{R}$ is called an exponential function.

  $f(x) = e^x$, $x \in \mathbb{R}$ is called **the** exponential function.

> $a^x > 0$ for all real values of $x$, so the range of an exponential is all positive real numbers.

The function $f(x) = e^x$ is the only function that is unchanged when it is differentiated.

Summing up:

**for any value of $a$ ($a > 0$), $f(x) = a^x$, $x \in \mathbb{R}$ is *an* exponential function,**

**for the base e (e $\approx$ 2.718), $f(x) = e^x$, $x \in \mathbb{R}$ is *the* exponential function.**

$$\frac{d}{dx}(e^x) = e^x$$

The following diagrams show sketches of $y = e^x$ and of some simple variations.

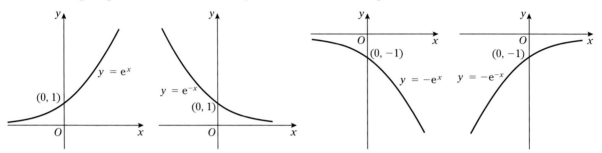

Powers of e, such as $e^2$, $e^{0.15}$, can be found from a calculator using the button marked '$e^x$'.

---

## Examples 3a

1  Find the coordinates of the stationary point on the curve $y = e^x - x$, and determine its type. Sketch the curve showing the stationary point clearly.

$$y = e^x - x \quad \Rightarrow \quad \frac{dy}{dx} = e^x - 1$$

At a stationary point, $\dfrac{dy}{dx} = 0$ therefore $e^x - 1 = 0$

i.e.      $e^x = 1 \quad \Rightarrow \quad x = 0$

When      $x = 0$, $y = e^0 - 0 = 1$

Therefore $(0, 1)$ is a stationary point.

$$\frac{d^2y}{dx^2} = e^x \text{ and this is positive when } x = 0$$

Therefore $(0, 1)$ is a minimum point.

This curve is made up from separate sketches of $y = e^x$ and $y = -x$ by adding their ordinates.

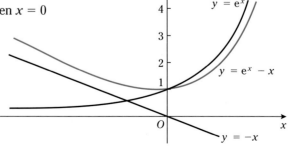

2  Find the derivative of $e^{(3x - 5)}$

When $y = e^{(3x - 5)}$, using $\dfrac{dy}{dx} = \dfrac{dy}{du} \times \dfrac{du}{dx}$ where $u = 3x - 5$ gives

$$\dfrac{dy}{dx} = (3) \times e^{(3x - 5)}$$

$\therefore$     the derivative of $e^{(3x - 5)}$ is $3e^{(3x - 5)}$

## Exercise 3a

1  Find the value of

  (a)  $e^2$            (b)  $e^{-1}$

  (c)  $e^{1.5}$          (d)  $e^{-0.3}$

  (e)  $e^3$            (f)  $e^{1.8}$

  (g)  $e^{-2}$          (h)  $e^{0.05}$

2  Write down the derivative of

  (a)  $2e^x$          (b)  $x^2 - e^x$

  (c)  $e^x$            (d)  $e^{(2 - 5x)}$

  (e)  $3e^{(2x - 4)}$     (f)  $2e^{(2 - x)}$

  (g)  $e^{(2x^2)}$        (h)  $3x^2 + e^{(x^3)}$

In questions 3 to 5 find the gradient of each curve at the specified value of $x$.

3  $y = e^x - 2x$ where $x = 2$

4  $y = x^2 + 2e^x$ where $x = 1$

5  $y = e^x - 3x^3$ where $x = 0$

6  Find the value of $x$ at which the function $e^x - 3x$ has a stationary value.

7  Sketch the given curve.

  (a)  $y = 1 - e^x$        (b)  $y = e^x + 1$

  (c)  $y = 1 - e^{-x}$      (d)  $y = 1 + e^{-x}$

## NATURAL LOGARITHMS

When e is used as the base for logarithms, they are called *natural logarithms*.
To avoid confusion with other bases, $\log_e x$ is written as $\ln x$.

**$\ln x$ means $\log_e x$**

**so**     **$\ln x = y$**    $\Leftrightarrow$    **$e^y = x$**

Natural logarithms are also called Naperian logarithms. You can evaluate a natural logarithm by using the 'ln' button on a calculator.

## Examples 3b

1  Separate $\ln(\tan x)$ into two terms.

$$\ln(\tan x) = \ln\left(\frac{\sin x}{\cos x}\right)$$

$$= \ln \sin x - \ln \cos x$$

2  Express $4 \ln(x + 1) - \frac{1}{2} \ln x$ as a single logarithm.

$$4 \ln(x + 1) - \frac{1}{2} \ln x = \ln(x + 1)^4 - \ln \sqrt{x}$$

$$= \ln \frac{(x + 1)^4}{\sqrt{x}}$$

## Exercise 3b

1  Convert each equation to logarithmic form.

   (a)  $e^x = 4$          (b)  $e^2 = y$

   (c)  $e^a = b$

2  Convert each equation to index form.

   (a)  $\ln x = 4$          (b)  $\ln 0.5 = a$

   (c)  $\ln a = b$

3  Use a calculator to evaluate, correct to 3 significant figures

   (a)  $\ln 3$          (b)  $\ln 2.4$

   (c)  $\ln 0.201$          (d)  $\ln 17.3$

4  Evaluate

   (a)  $\ln e$     (b)  $\ln e^2$     (c)  $\ln 1$

5  Express as a sum or difference of logarithms or as a product

   (a)  $\ln 5x$          (b)  $\ln 5x^2$

   (c)  $\ln 3(x + 1)$          (d)  $\ln \dfrac{x + 1}{x}$

   (e)  $\ln \dfrac{2x - 1}{x}$          (f)  $\ln xy^2$

   (g)  $\ln \sqrt{x + 1}$          (h)  $\ln x(x + 4)$

   (i)  $\ln (x^2 - 1)$          (j)  $\ln x^2(x + y)$

   (k)  $\ln ex$          (l)  $\ln e^2x(x - e)$

   (m)  $\ln \dfrac{x^2}{x + 1}$          (n)  $\ln (a^2 - b^2)$

   (o)  $\ln \tan x$

6  Express as a single logarithm

   (a)  $\ln 2 + \ln x$

   (b)  $\ln 3 - \ln x$

   (c)  $2 \ln x - \ln 4$

   (d)  $\ln x - 2 \ln (1 - x)$

   (e)  $1 - \ln x$

   (f)  $2 + \ln x$

   (g)  $2 \ln x - \frac{1}{2} \ln (x - 1)$

   (h)  $\ln \cos x - \ln \sin x$

   (i)  $1 + \ln x$

   (j)  $\frac{2}{3} \ln (x - 1)$

7  Solve the following equations for $x$.

   (a)  $e^x = 8.2$

   (b)  $e^{2x} + e^x - 2 = 0$          (Hint: use $e^{2x} = (e^x)^2$)

   (c)  $e^{2x - 1} = 3$ (d)  $e^{4x} + e^x = 0$

   (d)  $\ln 2 + 2 \ln x = \ln (x + 3)$

   (e)  $\ln 4 - 2 \ln (x + 1) = \ln e$

**P3 8**  Solve the equation $\ln (e^x - 1) = 1$

**P3 9**  Solve the equation $-3 + \ln x = \ln (x - 3)$

## THE LOGARITHMIC FUNCTION

$\ln x$ has a real value for every positive value of $x$. Therefore $x \mapsto \ln x$ is a function for $x > 0, x \in \mathbb{R}$ and it is called the logarithmic function.

When $y = e^x$, then $\ln y = x$,
so $\ln y$ maps values of $e^x$ back to $x$.

**Therefore the logarithmic function is the inverse of the exponential function.**

When $y = e^x$, interchanging $x$ and $y$ gives $x = e^y$ which can be written as $y = \ln x$

It follows that the curve $y = \ln x$ is the reflection of the curve $y = e^x$ in the line $y = x$

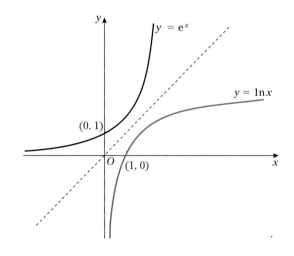

## THE DERIVATIVE OF ln $x$

We know that $y = \ln x \iff x = e^y$ and we also know how to differentiate the exponential function.

So a relationship between $\dfrac{d}{dx}(y)$ and $\dfrac{d}{dy}(x)$ is needed.

When $y = f(x)$ where $f(x)$ is any function of $x$,

$$\frac{dy}{dx} = \lim_{\delta x \to 0} \frac{\delta y}{\delta x} = \lim_{\delta x \to 0} \left( \frac{1}{\dfrac{\delta x}{\delta y}} \right)$$

Now $\delta y \to 0$ as $\delta x \to 0$,     $\therefore$    $\dfrac{dy}{dx} = \lim_{\delta y \to 0} \left( \dfrac{1}{\dfrac{\delta x}{\delta y}} \right)$

$$\frac{dy}{dx} = \frac{1}{\dfrac{dx}{dy}}$$

This relationship can be used to find the derivative of *any* function if the derivative of its inverse is known. We now apply it to differentiate ln $x$.

$$y = \ln x \qquad \iff \qquad x = e^y$$

Differentiating $e^y$ with respect to $y$ gives $\dfrac{dx}{dy} = e^y = x$

Therefore     $\dfrac{dy}{dx} = \dfrac{1}{\dfrac{dx}{dy}} = \dfrac{1}{x}$

i.e.
$$\frac{d}{dx} \ln x = \frac{1}{x}$$

This result can be used to differentiate many log functions if they are first simplified by applying the laws given on page 10.

---

### Examples 3c

1  Find the derivative of ln $(2x)$

$$f(x) = \ln(2x) = \ln 2 + \ln x$$

$$\frac{d}{dx}\{f(x)\} = 0 + \frac{1}{x} \qquad \text{ln 2 is a number}$$

The derivative of ln $(2x)$ is $\dfrac{1}{x}$.

2  Find the derivative of     (a)  $\ln\left(\dfrac{1}{x^3}\right)$     (b)  $\ln(4\sqrt{x})$

(a)     $f(x) = \ln\left(\dfrac{1}{x^3}\right) = \ln(x^{-3}) = -3 \ln x$

$$\frac{d}{dx}\{f(x)\} = \frac{d}{dx}\{-3\ln x\} = \frac{-3}{x}$$

(b)     $f(x) = \ln(4\sqrt{x}) = \ln 4 + \ln(\sqrt{x}) = \ln 4 + \frac{1}{2}\ln x$

$$\frac{d}{dx}\{f(x)\} = \frac{d}{dx}(\ln 4) + \frac{d}{dx}\left(\frac{1}{2}\ln x\right)$$

$$= 0 + \frac{\frac{1}{2}}{x} = \frac{1}{2x}$$

**3** Find the derivative of $\ln(x^2 + 2)$

When $y = \ln(x^2 + 2)$, and $u = x^2 + 2$, then $y = \ln u$     $\ln(x^2 + 2)$ cannot be separated into two logarithms so we use the chain rule.

$$\frac{dy}{dx} = \frac{dy}{du} \times \frac{du}{dx} = \frac{1}{u} \times (2x)$$

$$= \frac{2x}{x^2 + 2}$$

## Exercise 3c

**1** Write down the derivative of each of the following functions.

(a)  $\ln x^3$

(b)  $\ln(3x)$

(c)  $\ln(x^{-2})$

(d)  $\ln\left(\frac{3}{\sqrt{x}}\right)$

(e)  $\ln\left(\frac{1}{x^5}\right)$

(f)  $\ln\left(2x^{\frac{1}{2}}\right)$

(g)  $\ln\left(x^{-\frac{3}{2}}\right)$

(h)  $\ln\left(\frac{x^3}{\sqrt{x}}\right)$

**2** Locate the stationary points on each curve.

(a)  $y = \ln x - x$

(b)  $y = x^3 - 2\ln x^3$

(c)  $y = \ln x - \sqrt{x}$

**3** Sketch each of the following curves.

(a)  $y = -\ln x$

(b)  $y = \ln(-x)$

(c)  $y = 2 + \ln x$

(d)  $y = \ln x^2$

**4** Find the derivative of each of the following functions.

(a)  $\ln(3x - 1)$

(b)  $\ln(4x + 3)$

(c)  $\ln(x^2 + 1)$

(d)  $\ln(x^2 + 2x)$

(e)  $\ln(x^3 + 2)$

**To summarise,**

When $y = e^x$, $\frac{dy}{dx} = e^x$

and when $y = e^{f(x)}$, using $\frac{dy}{dx} = \frac{dy}{du} \times \frac{du}{dx}$ with $u = f(x)$ gives $\frac{dy}{dx} = e^u \times f'(x)$

$\therefore$     $\frac{dy}{dx} = f'(x)e^{f(x)}$

When $y = \ln x$, $\frac{dy}{dx} = \frac{1}{x}$

and when $y = \ln f(x)$, using $\frac{dy}{dx} = \frac{dy}{du} \times \frac{du}{dx}$ with $u = f(x)$ gives $\frac{dy}{dx} = \frac{1}{u} \times f'(x)$

$\therefore$     $\frac{dy}{dx} = \frac{f'(x)}{f(x)}$

These results are quotable, so the derivative of $y = e^{f(x)}$ and $y = \ln f(x)$ can be written down directly,

for example given $y = e^{(3x - 2)}$ then $\frac{dy}{dx} = 3e^{(3x - 2)}$

and given $y = \ln(x^2 + 1)$ then $\frac{dy}{dx} = \frac{2x}{x^2 + 1}$

## Mixed exercise 3d

1  Express $y$ in terms of $x$ when
   $\ln y - \ln (y + 1) = \ln x$

2  By sketching two graphs, show that $e^x = x^2$
   has only one root.

3  (a)  By sketching suitable graphs show that the
       equation $\ln x = 1 - x$ has only one root.

   (b)  Write down the value of this root.

4  Find the equation of the tangent to the curve
   $y = \ln (x^2)$ at the point where $x = 4$

5  Find the equation of the normal to the curve
   $y = 3e^{2x}$ where $x = 1$

6  Given that $y = e^{2x} - \ln \sqrt{x}$ find $\dfrac{dy}{dx}$.

7  Write down the derivatives of the following
   functions.

   (a)  $e^{3x + 2}$

   (b)  $\ln (x^2 - 3)$

   (c)  $e^{(x^2 - 1)}$

   (d)  $\ln (x^2 - 1)$

   (e)  $\ln (x^3 + 2x + 1)$

## REDUCTION OF A RELATIONSHIP TO A LINEAR FORM

A relationship of the form $y = ax^n$ where $a$ is a constant can be reduced to a linear form by taking logarithms, as

$$y = ax^n \quad \Leftrightarrow \quad \ln y = n \ln x + \ln a$$

Comparing     $\ln y = n \ln x + \ln a$

with          $Y = mX + c$

shows that plotting values of $\ln y$ against values of $\ln x$ gives a straight line whose gradient is $n$ and whose intercept on the vertical axis is $\ln a$.

### Examples 3e

1  Two variable quantities $p$ and $q$ are related by the equation $p = aq^n$. Some values of $p$ and $q$ are found from an experiment.

   The graph shows the result of plotting $\ln p$ against $\ln q$.

   Estimate the values of $a$ and $n$.

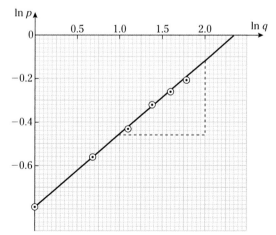

   Taking logs on both sides, $p = aq^n$ becomes $\ln p = n \ln q + \ln a$. Comparing with
   $y = mx + c$, $n$ is the gradient of the line and $\ln a$ is the intercept on the vertical axis.

   From the graph, the gradient of the line is 0.34, $\Rightarrow$ $n = 0.34$ and the intercept on the
   vertical axis is $-0.79$, so

   $\ln a = -0.79 \Rightarrow a = 0.5$

## Relationships of the form $y = ab^x$

A relationship of the form $y = ab^x$ where $a$ and $b$ are constant can be reduced to a linear relationship by taking logs, as

$$y = ab^x \quad \Leftrightarrow \quad \log y = x \log b + \log a$$

Comparing $\quad \log y = x \log b + \log a$

with $\qquad\qquad Y = mX + c$

shows that plotting values of $\log y$ against corresponding values of $x$ gives a straight line whose gradient is $\log b$ and whose intercept on the vertical axis is $\log a$.

---

### Examples 3e cont.

2  The variables $x$ and $y$ satisfy the relationship

$$2y + 10 = ab^{(x-3)}$$

The graph of $\ln (2y + 10)$ against $(x - 3)$ is a straight line as shown in the diagram.

Find the values of $a$ and $b$.

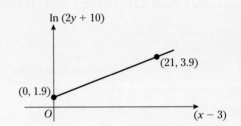

$$2y + 10 = ab^{(x-3)} \quad \Rightarrow \quad \ln (2y + 10) = (x - 3) \ln b + \ln a \qquad \text{This is the equation of the line in the diagram.}$$

Comparing with $Y = mX + c$ gives

$\quad \ln a$ is the intercept on the vertical axis,

$\therefore \quad \ln a = 1.9$ so $a = 6.69$ (3 s.f.)

and $\ln b$ is the gradient

$\therefore \quad \ln b = \dfrac{3.9 - 1.9}{21} = 0.09523...$ so $b = 1.10$ (3 s.f.)

---

### Exercise 3e

Throughout this exercise, $a$ and $b$ are constants.

1  The variables $x$ and $y$ satisfy the relation $ay = b^x$. The graph of $\ln y$ against $x$ is shown in the diagram.

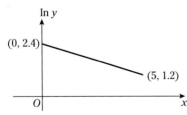

Find the values of $a$ and $b$.

2  The variables $s$ and $t$ satisfy the relation $s = ab^{-t}$. The graph of $\ln s$ plotted against $t$ gives a straight line that crosses the $s$-axis at

(0, 5) and crosses the $t$-axis at (10, 0). Find the values of $a$ and $b$.

3  The variables $x$ and $y$ satisfy the relation $y = a(x + 1)^b$

(a)  Take logarithms to show that plotting values of $\ln y$ against $\ln (x + 1)$ gives a straight line.

(b)  Given that $y = 20$ when $x = 2$ and that $y = 40$ when $x = 5$, find the gradient of the line.

4  The variables $x$ and $y$ satisfy the relation $3^x = 5^{y-2}$. Take natural logarithms to show that plotting values of $y$ against $x$ gives a straight line. Find the value of the intercept of this line on the vertical axis.

5 The variables $x$ and $y$ satisfy the relationship $3y + 5 = 7^{(x-1)}$. Explain why plotting $\ln(3y + 5)$ against $x$ gives a straight line. Find the coordinates of the point where the line crosses the $x$-axis.

P3 6 The variables $x$ and $y$ satisfy the relationship $y - 2000 = ab^{-x}$

It is given that $y = 8800$ when $x = 2$ and $y = 6100$ when $x = 6$

(a) Find the values of $a$ and $b$.

(b) Explain why the graph of $\ln(y - 200)$ against $x$ is a straight line.

P3 7 The variables $p$ and $q$ satisfy the relationship $p = aq^2$

It is given that $p = 100$ when $q = 50$

(a) Find the value of $a$.

(b) Explain why the graph of $\ln p$ against $q$ is a straight line.

P3 8 Two variables $s$ and $t$ are related by a law of the form $s = ke^{-nt}$ where $k$ and $n$ are constants. The values in the table were obtained from an experiment.

| $t$ | 1 | 1.5 | 2 | 2.5 | 3 |
|---|---|---|---|---|---|
| $s$ | 1230 | 590 | 260 | 140 | 60 |

(a) Show how the relationship between $s$ and $t$ can be reduced to a linear form.

(b) Draw a graph and use it to find the values of $k$ and $n$.

# 4 The modulus function

## After studying this chapter you should be able to

- understand the meaning of $|x|$, and use relations such as $|a| = |b| \Leftrightarrow a^2 = b^2$ and $|x - a| < b \Leftrightarrow a - b < x < a + b$ in the course of solving equations and inequalities.

## THE MODULUS OF $x$

The modulus of $x$ is written as $|x|$ and it means the positive value of $x$ whether $x$ itself is positive or negative, e.g. $|2| = 2$ and $|-2| = 2$.

Therefore the graph of $y = |x|$ can be found from the graph of $y = x$ by changing the part of the graph for which $y$ is negative to the equivalent positive values, i.e. by reflecting the part of the graph where $y$ is negative in the $x$-axis.

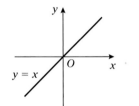 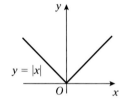

Hence, for $f : x \mapsto |x|$, $x \in \mathbb{R}$ we can write

$$\begin{cases} |x| = x \text{ for } x \geqslant 0 \\ |x| = -x \text{ for } x < 0 \end{cases}$$

## THE MODULUS OF A FUNCTION

The graph of any curve $C_1$ whose equation is $y = |f(x)|$ can be found from the curve $C_2$ with equation $y = f(x)$, by reflecting in the $x$-axis the parts of $C_2$ for which $f(x)$ is negative. The remaining sections of $C_1$ are not changed.

For example, to sketch $y = |(x - 1)(x - 2)|$ we start by sketching the curve $y = (x - 1)(x - 2)$. We then reflect in the $x$-axis the part of this curve which is below the $x$-axis.

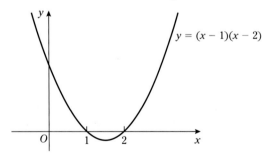 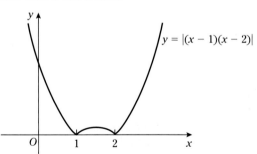

For any function f, the mapping $x \mapsto |f(x)|$ is also a function.

The diagrams show that the equation of the graph of $y = |(x - 1)(x - 2)|$ can also be written as

$$\begin{cases} y = (x - 1)(x - 2) \text{ for } x \leqslant 1 \text{ and } x \geqslant 2 \\ y = -(x - 1)(x - 2) \text{ for } 1 < x < 2 \end{cases}$$

For any curve of the form $y = |f(x)|$

$$\begin{cases} y = f(x) \text{ for values of } x \text{ where } f(x) \geqslant 0 \\ y = -f(x) \text{ for values of } x \text{ where } f(x) < 0 \end{cases}$$

### Example 4a

Sketch the curve $y = \left|1 - \tfrac{1}{2}x\right|$ and write the equations in non-modulus form of each part on the sketch.

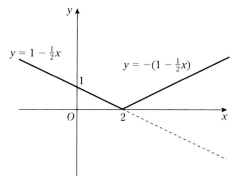

Start with a sketch of $y = 1 - \tfrac{1}{2}x$.

Then reflect the part below the $x$-axis in the $x$-axis.

### Exercise 4a

Sketch the following curves and write on the sketch the equation of each part of the curve in non-modulus form.

1  $y = |2x - 1|$

2  $y = |x - 4|$

3  $y = |x + 1|$

4  $y = |1 - 2x|$

5  $y = |3 - x|$

6  $y = |x(x - 1)(x - 2)|$

7  $y = |x^2 - 1|$

8  $y = |x^2 + 1|$

9  $y = |\sin x|$

10  $y = |\ln x|$

11  $y = |\cos x|$

12  $y = |x^2 - x - 20|$

## INTERSECTION

To find the points of intersection between two graphs whose equations involve a modulus, first sketch the graphs to locate the points roughly. Then we can identify the equations in non-modulus form for each part of the graph. When these equations are written on the sketch, the correct pair of equations for solving simultaneously can be seen.

Alternatively we can use the following fact:

We know that     $|a| = a$ for $a \geqslant 0$ and $|a| = -a$ for $a < 0$,

and that          $|b| = b$ for $b \geqslant 0$ and $|b| = -b$ for $b < 0$.

When $|a| = |b|$ then $a = \pm b$ so $a^2 = b^2$

and conversely when $a^2 = b^2$, $a = \pm b$ so $|a| = |b|$

We write this as $|a| = |b| \Leftrightarrow a^2 = b^2$

### Examples 4b

**1** Find the values of $x$ where the graph of $y = |x + 2|$ intersects the graph $y = |1 - 2x|$

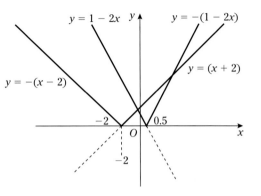

Using sketch graphs:

There are two points of intersection, one where $-(1 - 2x) = (x + 2) \Rightarrow x = 3$

and one where $1 - 2x = x + 2 \Rightarrow x = -\frac{1}{3}$

Alternatively using algebra:

$$|x + 2| = |1 - 2x| \Rightarrow (x + 2)^2 = (1 - 2x)^2$$

$$\therefore \qquad x^2 + 4x + 4 = 4x^2 - 4x + 1$$

$$\Rightarrow \qquad 3x^2 - 8x - 3 = 0$$

$$\Rightarrow \qquad (3x + 1)(x - 3) = 0$$

$$\therefore \qquad x = -\frac{1}{3} \text{ or } x = 3$$

Check:

when $x = -\frac{1}{3}$, $|x + 2| = 1\frac{2}{3}$ and $|1 - 2x| = 1\frac{2}{3}$, so $x = -\frac{1}{3}$ is a solution.

when $x = 3$, $|x + 2| = 5$ and $|1 - 2x| = 5$, so $x = 3$ is also a solution.

It is essential that the values of $x$ found by this method are checked, because squaring can sometimes introduce values of $x$ that are not solutions.

## SOLVING EQUATIONS INVOLVING MODULUS SIGNS

We can solve an equation such as $|2x - 1| = 3x$ by sketching the graphs of $y = |2x - 1|$ and $y = 3x$ and finding their points of intersection as illustrated above.

Alternatively we can use the fact that when $\quad |f(x)| = g(x)$

then $$f(x) = g(x) \quad \text{and} \quad -f(x) = g(x)$$

The next worked example shows how this method can be used.

### Examples 4b cont.

**2** Solve the equation $|2x - 1| = 3x$

Using $\quad f(x) = g(x) \quad$ and $\quad -f(x) = g(x)$

gives $\quad 2x - 1 = 3x \quad$ and $\quad 2x - 1 = -3x$

so $\quad x = -1$ (check $|2x - 1| = 3$ and $3x = -3$, so $x = -1$ is not a solution)

or $\quad x = \frac{1}{5}$ (check $|2x - 1| = \frac{3}{5}$ and $3x = \frac{3}{5}$, so $x = \frac{1}{5}$ is a solution)

Therefore $x = \frac{1}{5}$ is the only solution.

## Exercise 4b

Find the points of intersection of the graphs.

**1**  $y = |x - 1|$ and $y = |2x|$

**2**  $y = x$ and $y = |1 - 2x|$

**3**  $y = \left|\frac{1}{2}x\right|$ and $y = 3 + 2x$

**4**  $y = |2x - 5|$ and $y = |x|$

**5**  $y = |x - 4|$ and $y = 2x + 1$

Solve the following equations.

**6**  $|x - 1| = x$

**7**  $|3x - 2| = 4$

**8**  $|2 - x| = 6$

**9**  $|x^2 - 1| = 3x - 1$

**10**  $|x + 1| = |4x - 3|$

**11**  $|1 - x| = \frac{1}{2}x$

**P3 12**  $|2 - x^2| + 2x + 1 = 0$

**P3 13**  $3 - x = |x + 2|$

**P3 14**  $|x^2 - 6| = x$

**P3 15**  $|e^x - 2| = 2$

# INEQUALITIES

Many inequalities can be solved easily with the aid of sketch graphs.

The method used is the same as for equations.

Alternatively, for simple inequalities such as $|x - a| \leq b$ we can use the fact that, when $|x - a| \leq b$, the graph shows that $a - b \leq x \leq a + b$

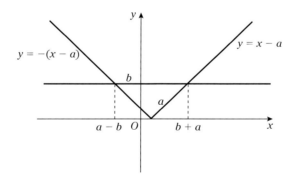

## Examples 4c

**1**  Solve the inequality $|3x - 2| < 4$

$|3x - 2| < 4$ so $3x - 2 < 4$ and $-(3x - 2) < 4$

giving $x < 2$ and $x > -\frac{2}{3}$,  i.e. $-\frac{2}{3} < x < 2$

Check by taking a value of $x$ between $-\frac{2}{3}$ and 2. When $x = 1$, $|3x - 2| = 1$ and $1 < 4$

**2  Solve the inequality $|3 - x| < |x|$**

From the sketch, $|3 - x| = |x|$
where $3 - x = x$, i.e. where $x = \frac{3}{2}$

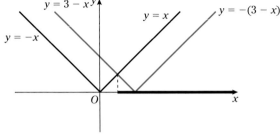

$\therefore \qquad |3 - x| < |x|$ for $x > \frac{3}{2}$

Alternatively, $\qquad |3 - x| = |x|$

$\Rightarrow \qquad (3 - x)^2 = x^2$

$\Rightarrow \qquad 9 - 6x = 0$, so $x = \frac{3}{2}$

This does not tell us whether $x > \frac{3}{2}$ or $x < \frac{3}{2}$ so choose value of $x < \frac{3}{2}$ and test it.

When $x = 1$, $|3 - x| = 2$ and $|x| = 1$, so $|3 - x| > |x|$ when $x < \frac{3}{2}$

therefore $|3 - x| < |x|$ when $x > \frac{3}{2}$

## Exercise 4c

Solve the following inequalities.

**1**  $|x - 1| > 3$

**2**  $|2x + 1| < 5$

**3**  $|4x - 1| > 3$

**4**  $|x - 1| < |x + 2|$

**5**  $|x - 3| < |2x - 3|$

**6**  $|x| < |2x - 3|$

**7**  $|2 - x| < |x|$

**8**  $|2 - x| > |2x - 1|$

**9**  $|2 - x| > |2x|$

**10**  $|4x - 1| > |2 - 2x|$

**11**  $|2 - x| < |3x|$

**12**  $|3 - 2x| > |2 - x|$

[P3]**13**  $|x + 1| < 2x$

[P3]**14**  $e^x < |x - 1|$

[P3]**15**  $|1 - x^2| < 2x + 1$

[P3]**16**  $|3^x - 2| < 1$

[P3]**17**  $|x| + 2 < 3x$

[P3]**18**  $|x^2 - x - 5| > 3x$

# 5 Diffferentiation 1

*After studying this chapter you should be able to*

- differentiate products and quotients.

## DIFFERENTIATING A PRODUCT

When $y = uv$ where $u$ and $v$ are both functions of $x$, e.g. $y = x^2(x^4 - 1)$

you could think that $\dfrac{dy}{dx}$ is given by $\left(\dfrac{du}{dx}\right)\left(\dfrac{dv}{dx}\right)$.

But this is *not* true, as is shown by an example such as $y = (x^2)(x^3)$ where, because $y = x^5$, we know that $\dfrac{dy}{dx} = 5x^4$ which is *not* equal to $(2x)(3x^2)$.

When $y = uv$ where $u = f(x)$ and $v = g(x)$, then when $x$ increases by a small amount $\delta x$, there are small increases of $\delta u$, $\delta v$ and $\delta y$ in the values of $u$, $v$ and $y$ respectively.

$\therefore \qquad y + \delta y = (u + \delta u)(v + \delta v)$

$\qquad\qquad\qquad = uv + u\delta v + v\delta u + \delta u \delta v$

But $y = uv$

$\therefore \qquad\qquad \delta y = u\delta v + v\delta u + \delta u \delta v$

$\Rightarrow \qquad\qquad \dfrac{\delta y}{\delta x} = u\dfrac{\delta v}{\delta x} + v\dfrac{\delta u}{\delta x} + \delta u\dfrac{\delta v}{\delta x}$

As $\delta x \to 0$, $\dfrac{\delta v}{\delta x} \to \dfrac{dv}{dx}$, $\dfrac{\delta u}{\delta x} \to \dfrac{du}{dx}$ and $\delta u \to 0$

Therefore $\qquad \dfrac{dy}{dx} = \lim_{\delta x \to 0} \dfrac{\delta y}{\delta x}$

$\qquad\qquad\qquad = u\dfrac{dv}{dx} + v\dfrac{du}{dx} + 0$

i.e. $\qquad \dfrac{d}{dx}(uv) = v\dfrac{du}{dx} + u\dfrac{dv}{dx}$

---

### Examples 5a

1 Differentiate $x^2 e^x$ with respect to $x$.

Using $u = x^2$ and $v = e^x$

$\qquad \dfrac{du}{dx} = 2x$ and $\dfrac{dv}{dx} = e^x$

then $\dfrac{d}{dx}(uv) = v\dfrac{du}{dx} + u\dfrac{dv}{dx}$ gives

$\dfrac{dy}{dx} = (e^x)(2x) + (x^2)(e^x) = e^x(x^2 + 2x)$

**2** Differentiate $(x + 1)^3(2x - 5)$ with respect to $x$.

Using $u = (x + 1)^3$ and $v = (2x - 5)$

$$\frac{du}{dx} = 3(x + 1)^2 \text{ and } \frac{dv}{dx} = 2$$

then $\frac{d}{dx}(uv) = v\frac{du}{dx} + u\frac{dv}{dx}$ gives

$$\frac{dy}{dx} = (2x - 5) \times 3(x + 1)^2 + (x + 1)^3 \times 2$$

$$= (x + 1)^2[3(2x - 5) + 2(x + 1)]$$

$$= (x + 1)^2(8x - 13)$$

## Exercise 5a

Differentiate each function with respect to $x$.

1  $x(x - 3)^2$

2  $x\sqrt{x - 6}$

3  $xe^{2x}$

4  $x(2x + 3)^3$

5  $x \ln x$

6  $(x - 1)e^x$

7  $(x + 1) \ln 2x$

8  $(x + 1)\sqrt{x}$

9  $x^2 e^{x^2}$

10 $x^3 (x - 1)^2$

11 $x(x + 3)^{-1}$

12 $x^3 \ln (x - 1)$

13 $(x^2 - 1)e^x$

14 $x \ln (x^2 - 3x)$

## DIFFERENTIATING A QUOTIENT

To differentiate a function of the form $\frac{u}{v}$, where $u$ and $v$ are both functions of $x$, we can rewrite the function as $uv^{-1}$ and differentiate it as a product. We can also use the formula derived below.

When a function is of the form $\frac{u}{v}$, where $u$ and $v$ are both functions of $x$, a small increase of $\delta x$ in the value of $x$ causes small increases of $\delta u$ and $\delta v$ in the values of $u$ and $v$ respectively.
Then, as $\delta x \to 0$, $\delta u$ and $\delta v$ also tend to zero.

If $y = \frac{u}{v}$ then $y + \delta y = \frac{(u + \delta u)}{(v + \delta v)}$

$\therefore \qquad \delta y = \frac{u + \delta u}{v + \delta v} - \frac{u}{v} = \frac{v\delta u - u\delta v}{v(v + \delta v)}$

$\therefore \qquad \frac{\delta y}{\delta x} = \frac{\left(v\dfrac{\delta u}{\delta x} - u\dfrac{\delta v}{\delta x}\right)}{v(v + \delta v)}$

$\Rightarrow \qquad \frac{dy}{dx} = \lim_{\delta x \to 0} \frac{\delta y}{\delta x} = \frac{\left(v\dfrac{du}{dx} - u\dfrac{dv}{dx}\right)}{v^2}$

i.e.

$$\frac{dy}{dx} = \frac{v\dfrac{du}{dx} - u\dfrac{dv}{dx}}{v^2}$$

**Example 5b**

Given $y = \dfrac{e^x}{(x+1)^2}$ find $\dfrac{dy}{dx}$

Using $u = e^x$ and $v = (x+1)^2$

gives $\dfrac{du}{dx} = e^x$ and $\dfrac{dv}{dx} = 2(x+1)$

then $\dfrac{dy}{dx} = \dfrac{v\dfrac{du}{dx} - u\dfrac{dv}{dx}}{v^2}$ gives

$$\frac{dy}{dx} = \frac{(x+1)^2 e^x - e^x[2(x+1)]}{(x+1)^4} = \frac{e^x(x^2-1)}{(x+1)^4} = \frac{e^x(x+1)(x-1)}{(x+1)^4} = \frac{(x-1)e^x}{(x+1)^3}$$

**Exercise 5b**

Use the quotient formula to differentiate each of the following functions with respect to $x$.

1  $\dfrac{(x-3)^2}{x}$

2  $\dfrac{x^2}{(x+3)}$

3  $\dfrac{(4-x)}{x^2}$

4  $\dfrac{(x+1)^2}{x^3}$

5  $\dfrac{4x}{(1-x)^3}$

6  $\dfrac{2x^2}{(x-2)}$

7  $\dfrac{e^x}{x-1}$

8  $\dfrac{\ln x}{x}$

9  $\dfrac{e^{2x}}{(x+1)^2}$

10  $\dfrac{x}{e^x}$

11  $\dfrac{e^x}{x^2}$

12  $\dfrac{(\ln x)}{x^3}$

13  $\dfrac{e^x}{x^2-1}$

14  $\dfrac{e^x}{e^x - e^{-x}}$

15  $\dfrac{1}{\ln x}$

# IDENTIFYING THE CATEGORY OF A FUNCTION

Before any formula can be used to differentiate a given function, it is important to recognise which category the function belongs to, e.g. is it a product or a quotient or a composite function. Remember that

(a)  a product has two separate parts, each being an independent function of $x$; you can actually put brackets round the parts and each is complete in itself, e.g. $(e^x)(\sin x)$.

(b)  whereas if *one operation* is carried out *on another function of $x$* we have a composite function e.g. $e^{\sin x}$ is an exponential function of a trigonometric function and it cannot be separated into independent parts.

Remember also that sometimes a fraction can be expressed more simply as a product,

e.g.    $\dfrac{\sin x}{e^x}$ can be written as $e^{-x}\sin x$.

## Mixed exercise 5

In each case first identify the type of function and then use the appropriate method to find its derivative. Some of the functions can be differentiated by using a basic rule, so do not assume that a formula is always needed.

1 $x\sqrt{x + 1}$

2 $(x^2 - 8)^3$

3 $\dfrac{x}{x^2 + 1}$

4 $\sqrt[3]{2 - x^4}$

5 $\dfrac{x^2 + 1}{x^2 + 2}$

6 $x^2 (\sqrt{x} - 2)$

7 $(x^2 - 2)^3$

8 $\sqrt{x - x^2}$

9 $\dfrac{x}{\sqrt{x} + 1}$

10 $x^2\sqrt{x - 2}$

11 $\dfrac{\sqrt{x + 1}}{x^2}$

12 $(x^4 + x^2)^3$

13 $\sqrt{x^2 - 8}$

14 $x^3(x^2 - 6)$

15 $(x^2 - 6)^3$

16 $\dfrac{x}{x^2 - 6}$

17 $(x^4 + 3)^{-2}$

18 $\sqrt{x}(2 - x)^3$

19 $\dfrac{\sqrt{x}}{(2 - x)^3}$

20 $(x - 1)(x - 2)^2$

21 $(2x^3 + 4)^5$

22 $x \ln x$

23 $(4x - 1)^{\frac{2}{3}}$

24 $\dfrac{e^x}{x - 1}$

25 $\dfrac{\sqrt{1 + x^3}}{x^2}$

26 $\dfrac{\ln x}{\ln (x - 1)}$

27 $(\ln x)^2$

28 $\dfrac{(1 + 2x^2)}{1 + x^2}$

29 $e^{-\frac{2}{x}}$

30 $\ln (1 - e^x)$

31 $e^{3x}x^3$

32 $\dfrac{2x}{(2x - 1)(x - 3)}$

33 $\dfrac{e^{\frac{x}{2}}}{x^5}$

34 $\ln \left[ \dfrac{x^2}{(x + 3)} \right]$

35 $\ln \left[ 4x^3(x + 3)^2 \right]$

36 $(\ln x)^4$

37 $\dfrac{(x + 3)^3}{x^2 + 2}$

38 $\sqrt{e^x - x}$

39 $4 \ln (x^2 + 1)$

Find and simplify $\dfrac{dy}{dx}$ and hence find $\dfrac{d^2y}{dx^2}$

40 $y = \dfrac{1 + 2x}{1 - 2x}$

41 $y = \ln \dfrac{x}{x + 1}$

42 $y = \dfrac{e^x}{e^x - 4}$

# Summary 1

### The remainder theorem

When a polynomial, f(x), is divided by $(x - a)$ the remainder is equal to f(a).

## FACTOR THEOREM

$(x - a)$ is a factor of a polynomial if and only if the polynomial is equal to zero when $a$ is substituted for $x$.

## LOGARITHMS

$$\log_a b = c \quad \Leftrightarrow \quad a^c = b$$

$$\log_a b + \log_a c = \log_a bc$$

$$\log_a b - \log_a c = \log_a \frac{b}{c}$$

$$\log_a b^n = n \log_a b$$

### Reduction of relationships to linear form

When a non-linear law, containing two unknown constants, connects two variables, the relationship can often be reduced to linear form. The aim is to produce an equation in which one term is constant and another term does not contain a constant. The law can then be expressed in the form

$$Y = mX + C$$

Some common conversions are

$y = ax^n$; take logs: $\ln y = \ln a + n \ln x$; use $Y = \ln y$ and $X = \ln x$

$y = ab^x$; take logs: $\ln y = \ln a + x \ln b$; use $Y = \ln y$ and $X = x$

## DIFFERENTIATION OF PRODUCTS AND QUOTIENTS

If $u$ and $v$ are both functions of $x$ then

$$y = uv \quad \Rightarrow \quad \frac{dy}{dx} = v\frac{du}{dx} + u\frac{dv}{dx}$$

$$y = \frac{u}{v} \quad \Rightarrow \quad \frac{dy}{dx} = \frac{\left(v\frac{du}{dx} - u\frac{dv}{dx}\right)}{v^2}$$

## FUNCTIONS

### Exponential functions

$a^x$, $x \in \mathbb{R}$ is an exponential function.

$e^x$, $x \in \mathbb{R}$ is *the* exponential function where

$e = 2.71828...$  and  $\frac{d}{dx}(e^x) = e^x$

## Logarithmic function

$\ln x$ means $\log_e x$ where $e = 2.71828...$

$\log_a x, x > 0, x \in \mathbb{R}$ is a logarithmic function.

$\log_e x = \ln x, x > 0, x \in \mathbb{R}$ is the natural logarithmic function and $\dfrac{d}{dx}(\ln x) = \dfrac{1}{x}$

The natural logarithmic function is the inverse of the exponential function, i.e.

$\quad f(x) = e^x \Rightarrow f^{-1}(x) = \ln x$

## Modulus functions

$|x|$ is the modulus function where $|x|$ is the positive numerical value of $x$, i.e. when $x = -3$, $|x| = 3$

The curve $y = |f(x)|$ is obtained from the curve $y = f(x)$ by reflecting in the $x$-axis the parts of the curve for which $f(x)$ is negative. The section(s) for which $f(x)$ is positive remain unchanged.

## Summary exercise 1

1  Solve the inequality $|x - 4| > |x + 1|$ [4]
Cambridge, Paper 2 Q1 J03

2  Given that $2^x = 5^y$, use logarithms to find the value of $\dfrac{x}{y}$ correct to 3 significant figures. [3]
Cambridge, Paper 2 Q1 J04

3  The cubic polynomial $2x^3 + ax^2 - 13x - 6$ is denoted by $f(x)$. It is given that $(x - 3)$ is a factor of $f(x)$.

(i) Find the value of $a$. [2]

(ii) When $a$ has this value, solve the equation $f(x) = 0$ [4]
Cambridge, Paper 2 Q3 J04

4  Solve the inequality $|x - 3| > |x + 2|$ [4]
Cambridge, Paper 2 Q1 J07

5  The variables $x$ and $y$ satisfy the relation $3^y = 4^{x+2}$

(i) By taking logarithms, show that the graph of $y$ against $x$ is a straight line. Find the exact value of the gradient of this line. [3]

(ii) Calculate the $x$-coordinate of the point of intersection of this line with the line $y = 2x$, giving your answer correct to 2 decimal places. [3]
Cambridge, Paper 2 Q2 J07

6  Solve the inequality $|3x - 1| < 2$ [3]
Cambridge, Paper 2 Q1 J08

7  Use logarithms to solve the equation $4^x = 2(3^x)$, giving your answer correct to 3 significant figures. [4]
Cambridge, Paper 2 Q2 J08

8  The polynomial $2x^3 + 7x^2 + ax + b$, where $a$ and $b$ are constants, is denoted by $p(x)$. It is given that $(x + 1)$ is a factor of $p(x)$, and that when $p(x)$ is divided by $(x + 2)$ the remainder is 5. Find the values of $a$ and $b$. [5]
Cambridge, Paper 2 Q4 J08

9  Given that $(1.25)^x = (2.5)^y$, use logarithms to find the value of $\dfrac{x}{y}$ correct to 3 significant figures. [3]
Cambridge, Paper 2 Q1 J09

10  Solve the inequality $|3x + 2| < |x|$ [4]
Cambridge, Paper 2 Q2 J09

11  The polynomial $x^3 + ax^2 + bx + 6$, where $a$ and $b$ are constants, is denoted by $p(x)$. It is given that $(x - 2)$ is a factor of $p(x)$, and that when $p(x)$ is divided by $(x - 1)$ the remainder is 4.

(i) Find the values of $a$ and $b$. [5]

(ii) When $a$ and $b$ have these values, find the other two linear factors of $p(x)$. [3]
Cambridge, Paper 2 Q6 J09

12    (i) Express $9^x$ in terms of $y$, where $y = 3^x$
                                                            [1]

      (ii) Hence solve the equation
              $2(9^x) - 7(3^x) + 3 = 0$
           expressing your answers for $x$ in terms
           of logarithms where appropriate.    [5]
                         Cambridge, Paper 2 Q3 N02

13    Given that $13^x = (2.8)^y$, use logarithms to
      show that $y = kx$ and find the value of $k$
      correct to 3 significant figures.    [3]
                         Cambridge, Paper 22 Q1 J10

14    Solve the inequality $|2x - 1| < |x + 4|$  [4]
                         Cambridge, Paper 22 Q3 J10

15    (i) Solve the inequality $|y - 5| < 1$    [2]
      (ii) Hence solve the inequality $|3^x - 5| < 1$,
           giving 3 significant figures in your
           answer.    [3]
                         Cambridge, Paper 2 Q3 N07

16

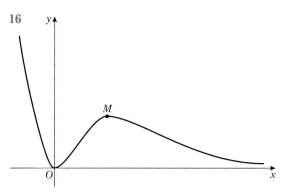

      The diagram shows the curve $y = x^2 e^{-x}$ and
      its maximum point $M$.
      (i) Find the $x$-coordinate of $M$.    [4]
      (ii) Show that the tangent to the curve at
           the point where $x = 1$ passes through
           the origin.    [3]
                 Cambridge, Paper 2 Q8 parts i and ii N07

17

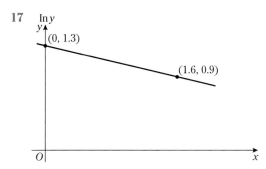

      The variables $x$ and $y$ satisfy the equation
      $y = A(b^{-x})$, where $A$ and $b$ are constants.

The graph of $\ln y$ against $x$ is a straight line
passing through the points $(0, 1.3)$ and
$(1.6, 0.9)$, as shown in the diagram.
Find the values of $A$ and $b$, correct to 2
decimal places.    [5]
                         Cambridge, Paper 2 Q3 N08

18    Find the exact coordinates of the point on
      the curve $y = xe^{-\frac{1}{2}x}$ at which $\dfrac{d^2y}{dx^2} = 0$    [7]

                         Cambridge, Paper 2 Q6 N08

19    The curve with equation $y = x \ln x$ has one
      stationary point.
      (i) Find the exact coordinates of this point,
          giving your answers in terms of e.    [5]
      (ii) Determine whether this point is a
           maximum or a minimum point.    [2]
                         Cambridge, Paper 2 Q6 N09

20    Solve the equation $\ln (3 - x^2) = 2 \ln x$,
      giving your answer correct to 3 significant
      figures.    [4]
                         Cambridge, Paper 21 Q2 N09

21

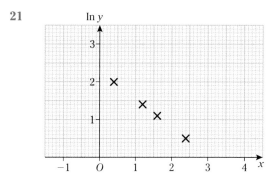

Two variable quantities $x$ and $y$ are related
by the equation

      $$y = k(a^{-x})$$

where $a$ and $k$ are constants. Four
pairs of values of $x$ and $y$ are measured
experimentally. The result of plotting $\ln y$
against $x$ is shown in the diagram.
Use the diagram to estimate the values
of $a$ and $k$.    [5]
                         Cambridge, Paper 2 Q2 N03

# 6 Trigonometic identities

*After studying this chapter you should be able to*

- understand the relationship of the secant, cosecant and cotangent functions to cosine, sine and tangent, and use properties and graphs of all six trigonometric functions for angles of any magnitude
- use trigonometrical identities for the simplification and exact evaluation of expressions and in the course of solving equations, and select an identity or identities appropriate to the context, showing familiarity in particular with the use of

    $\sec^2 \theta \equiv 1 + \tan^2 \theta$ and $\operatorname{cosec}^2 \theta \equiv 1 + \cot^2 \theta$

    the expansions of $\sin (A \pm B)$, $\cos (A \pm B)$ and $\tan (A \pm B)$,

    the formulae for $\sin 2A$, $\cos 2A$ and $\tan 2A$

    the expressions of $a \sin \theta + b \cos \theta$ in the forms $R \sin (\theta \pm \alpha)$ and $R \cos (\theta \pm \alpha)$

## THE RECIPROCAL TRIGONOMETRIC FUNCTIONS

The reciprocals of the three main trigonometric functions have their own names.

$$\frac{1}{\sin \theta} \equiv \operatorname{cosec} \theta, \qquad \frac{1}{\cos \theta} \equiv \sec \theta, \qquad \frac{1}{\tan \theta} \equiv \cot \theta$$

The names given above are abbreviations of cosecant, secant and cotangent respectively.

The graph of $f(\theta) = \operatorname{cosec} \theta$ is given below.

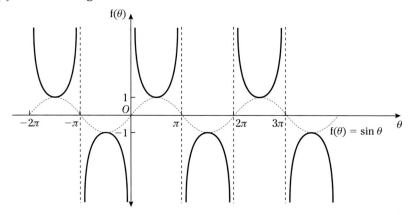

From this graph we can see that the cosec function is not continuous; it is undefined when $\theta$ is any integral multiple of $\pi$ (we would expect this because these are values of $\theta$ where $\sin \theta = 0$ and $\frac{1}{0}$ is undefined).

The pattern of the graph $f(\theta) = \sec \theta$ is similar to that of the cosec graph, as we would expect.

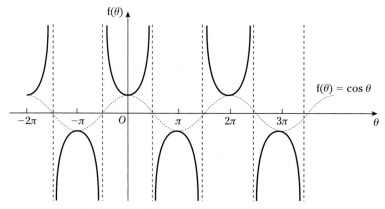

The graph of $f(\theta) = \cot \theta$ is given below.

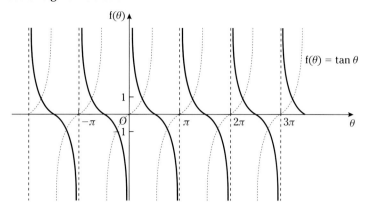

In this right-angled triangle,

$\tan \alpha = \dfrac{a}{b}$   and   $\cot \beta = \dfrac{a}{b} \left( \cot \beta = \dfrac{1}{\tan \beta} \right)$

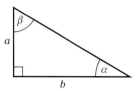

Now $\alpha + \beta = 90°$, i.e. $\alpha$ and $\beta$ are complementary angles. Hence the cotangent of an angle is equal to the tangent of its complement.

In fact, for *any* angle $\theta$,   $\cot \theta \equiv \tan \left( \tfrac{1}{2}\pi - \theta \right)$

We know that   $\tan \theta = \dfrac{\sin \theta}{\cos \theta}$   so   $\cot \theta = \dfrac{\cos \theta}{\sin \theta}$

---

### Example 6a

For $0 \leqslant \theta \leqslant 360°$, find the values of $\theta$ for which $\operatorname{cosec} \theta = -8$

$$\sin \theta = \frac{1}{\operatorname{cosec} \theta} = -\frac{1}{8} = -0.125$$

$\therefore$     from a calculator $\theta = -7.2°$

From the sketch, the required values of $\theta$ are
187.2° and 352.8°

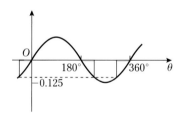

---

### Exercise 6a

1  Find, in the range $0 \leqslant \theta \leqslant 360°$, the values of $\theta$ for which

(a)  $\sec \theta = 2$    (b)  $\cot \theta = 0.6$

(c)  $\operatorname{cosec} \theta = 1.5$

2  In the range $-180° \leqslant \theta \leqslant 180°$ find the values of $\theta$ for which

(a)  $\cot \theta = 1.2$    (b)  $\sec \theta = -1.5$

(c)  $\operatorname{cosec} \theta = -2$

3  Given that $\tan \theta \equiv \dfrac{\sin \theta}{\cos \theta}$, write $\cot \theta$ in terms of $\sin \theta$ and $\cos \theta$.
Hence show that $\cot \theta - \cos \theta = 0$ can be written in the form $\cos \theta (1 - \sin \theta) = 0$, provided that $\sin \theta \neq 0$
Hence find the values in the range $-\pi \leqslant \theta \leqslant \pi$ for which $\cot \theta - \cos \theta = 0$

4  Find, in surd form if necessary, the values of

(a)  $\cot \tfrac{1}{4}\pi$    (b)  $\sec \tfrac{5}{4}\pi$    (c)  $\operatorname{cosec} \tfrac{11}{6}\pi$

## THE PYTHAGOREAN IDENTITIES

We know that $\cos^2 \theta + \sin^2 \theta = 1$ [1]

Using the identity $\tan \theta \equiv \dfrac{\sin \theta}{\cos \theta}$ we can write [1] in two other forms.

$[1] \div \cos^2 \theta \quad \Rightarrow \quad 1 + \dfrac{\sin^2 \theta}{\cos^2 \theta} \equiv \dfrac{1}{\cos^2 \theta}$

$\Rightarrow \qquad\qquad\qquad\qquad \mathbf{1 + \tan^2\, \theta \equiv \sec^2\, \theta}$

$[1] \div \sin^2 \theta \quad \Rightarrow \quad \dfrac{\cos^2 \theta}{\sin^2 \theta} + 1 \equiv \dfrac{1}{\sin^2 \theta}$

$\Rightarrow \qquad\qquad\qquad\qquad \mathbf{\cot^2\, \theta + 1 \equiv \mathrm{cosec}^2\, \theta}$

These identities can be used to

    simplify trigonometric expressions,

    eliminate trigonometric terms from pairs of equations,

    derive a variety of further trigonometric relationships,

    calculate other trigonometric ratios of any angle for which one trigonometric ratio is known,

    solve equations.

---

### Examples 6b

**1** Simplify $\dfrac{\sin \theta}{1 + \cot^2 \theta}$

$$\frac{\sin \theta}{1 + \cot^2 \theta} \equiv \frac{\sin \theta}{\mathrm{cosec}^2 \theta}$$

$$\equiv \sin^3 \theta$$

Using $1 + \cot^2 \theta \equiv \mathrm{cosec}^2 \theta$ and $\mathrm{cosec}\, \theta \equiv \dfrac{1}{\sin \theta}$

**2** Eliminate $\theta$ from the equations $x = \tan \theta$ and $3y = \cos \theta$

We need to use an identity that connects $\tan \theta$ and $\cos \theta$. There is not a direct identity but as $\sec \theta = \dfrac{1}{\cos \theta}$, we can use the identity $1 + \tan^2 \theta \equiv \sec^2 \theta$

$\cos \theta = 3y$ so $\sec \theta = \dfrac{1}{3y}$

Therefore $\qquad 1 + \tan^2 \theta \equiv \sec^2 \theta \Rightarrow 1 + x^2 = \left(\dfrac{1}{3y}\right)^2$

i.e. $\qquad 9y^2 + 9x^2y^2 = 1$

**3** Prove that $(1 - \cos A)(1 + \sec A) \equiv \sin A \tan A$

Because the relationship has yet to be proved, we must not assume its truth by using the complete identity in our working. The left- and right-hand sides must be isolated throughout the proof, preferably by working on only one of these sides. In general, start with the more complicated side. It often helps to express all ratios in terms of sine and/or cosine as, usually, these are easier to work with.

Consider the left-hand side:

$$(1 - \cos A)(1 + \sec A) \equiv 1 + \sec A - \cos A - \cos A \sec A$$

$$\equiv 1 + \sec A - \cos A - \cos A\left(\frac{1}{\cos A}\right)$$

$$\equiv \sec A - \cos A \equiv \frac{1}{\cos A} - \cos A$$

$$\equiv \frac{1 - \cos^2 A}{\cos A} \equiv \frac{\sin^2 A}{\cos A} \qquad\qquad \cos^2 A + \sin^2 A \equiv 1$$

$$\equiv \sin A\left[\frac{\sin A}{\cos A}\right] \equiv \sin A \tan A \equiv \text{right-hand side}$$

4  Solve the equation $\cos^2 \theta + \sec^2 \theta = 2$ for angles in the range $0 \leqslant \theta \leqslant 360^\circ$

$$\frac{\cos^4 \theta + 1}{\cos^2 \theta} = 2$$

> Equations are easier to solve when they contain just $\sin \theta$ or $\cos \theta$.
> We can use $\sec \theta = \dfrac{1}{\cos \theta}$ to give an equation containing just $\cos \theta$.

$$\Rightarrow \qquad \cos^4 \theta + 1 = 2\cos^2 \theta$$

provided $\cos \theta \neq 0$

$$\Rightarrow \qquad \cos^4 \theta - 2\cos^2 \theta + 1 = 0$$

$$\Rightarrow \qquad (\cos^2 \theta - 1)^2 = 0$$

so $\quad \cos^2 \theta = 1 \quad \Rightarrow \quad \cos \theta = \pm 1$

$\cos \theta = 1 \quad \Rightarrow \quad \theta = 0 \text{ or } 360^\circ, \qquad \cos \theta = -1 \quad \Rightarrow \quad \theta = 180^\circ$

5  Find the solution of the equation $\cot\left(\frac{1}{3}\theta - 90^\circ\right) = 1$, for which $0 \leqslant \theta \leqslant 540^\circ$

Using $\quad \frac{1}{3}\theta - 90^\circ = \phi$ gives

$$\cot\left(\frac{1}{3}\theta - 90^\circ\right) = \cot \phi$$

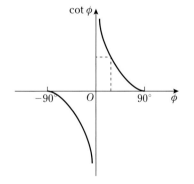

As $\theta$ is required in the range $0 \leqslant \theta \leqslant 540^\circ$ we must find
$\phi$ in the range $\frac{1}{3}(0) - 90^\circ \leqslant \phi \leqslant \frac{1}{3}(540^\circ) - 90^\circ$
i.e. $-90^\circ \leqslant \phi \leqslant 90^\circ$

The solution of the equation $\cot \phi = 1$ is

$$\phi = 45^\circ$$

But

$$\phi = \frac{1}{3}\theta - 90^\circ$$

so

$$\frac{1}{3}\theta - 90^\circ = 45^\circ$$

$$\Rightarrow \qquad \frac{1}{3}\theta = 135^\circ$$

i.e.

$$\theta = 405^\circ$$

## Exercise 6b

Simplify the following expressions.

**1**  $\dfrac{1 - \sec^2 A}{1 - \text{cosec}^2 A}$

**2**  $\dfrac{\sqrt{(1 + \tan^2 \theta)}}{\sqrt{(1 - \sin^2 \theta)}}$

**3**  $\dfrac{1}{\cos \theta \sqrt{(1 + \cot^2 \theta)}}$

**4**  $\dfrac{\sin \theta}{1 + \cot^2 \theta}$

Eliminate $\theta$ from the following pairs of equations.

**5**  $x = 4 \sec \theta$
    $y = 4 \tan \theta$

**6**  $x = a \, \text{cosec} \, \theta$
    $y = b \cot \theta$

**7**  $x = a \sec \theta$
    $y = b \sin \theta$

Prove the following identities.

**8**  $\cot \theta + \tan \theta \equiv \sec \theta \, \text{cosec} \, \theta$

**9**  $\dfrac{\cos A}{1 - \tan A} + \dfrac{\sin A}{1 - \cot A} \equiv \sin A + \cos A$

**10**  $\tan^2 \theta + \cot^2 \theta \equiv \sec^2 \theta + \text{cosec}^2 \theta - 2$

Solve the equations for angles in the range $0 \leqslant \theta \leqslant 360°$

**11**  $\cot^2 \theta = \text{cosec} \, \theta$

**12**  $\sec^2 \theta + \tan^2 \theta = 6$

**13**  $\tan \theta + \cot \theta = 2 \sec \theta$

**14**  $\tan \theta + 3 \cot \theta = 5 \sec \theta$

**15**  $4 \sec^2 \theta - 3 \tan \theta = 5$

**16**  $4 \cot^2 \theta + 12 \, \text{cosec} \, \theta + 1 = 0$

## COMPOUND ANGLES

It is easy to think that $\sin (A + B)$ is $\sin A + \sin B$. However, this is not true as can be seen from

$$\sin (45° + 45°) = \sin 90° = 1$$

whereas     $\sin 45° + \sin 45° = \tfrac{1}{2}\sqrt{2} + \tfrac{1}{2}\sqrt{2} \neq 1$

Also     $\cos (A + B)$ is NOT $\cos A + \cos B$

and     $\tan (A + B)$ is NOT $\tan A + \tan B$

The correct identity for $\sin (A + B)$ is $\sin (A + B) \equiv \sin A \cos B + \cos A \sin B$

This is proved geometrically, when $A$ and $B$ are both acute, from the diagram.

The right-angled triangles $OPQ$ and $OQR$ contain angles $A$ and $B$ as shown.

From the diagram, $\angle URQ = A$

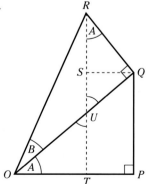

$$\sin (A + B) = \frac{TR}{OR} = \frac{TS + SR}{OR} = \frac{PQ + SR}{OR}$$

$$= \frac{PQ}{OQ} \times \frac{OQ}{OR} + \frac{SR}{QR} \times \frac{QR}{OR}$$

$\therefore$     $\sin (A + B) \equiv \sin A \cos B + \cos A \sin B$

This identity is valid for all angles. It can be adapted to give the full set of compound angle formulae. You are left to do this in the following exercise.

## Exercise 6c

1 In the identity
$\sin (A + B) \equiv \sin A \cos B + \cos A \sin B$,
replace $B$ by $-B$ to show that
$\sin (A - B) \equiv \sin A \cos B - \cos A \sin B$

2 In the identity
$\sin (A - B) \equiv \sin A \cos B - \cos A \sin B$,
replace $A$ by $\left(\frac{1}{2}\pi - A\right)$ to show that
$\cos (A + B) \equiv \cos A \cos B - \sin A \sin B$

3 In the identity
$\cos (A + B) \equiv \cos A \cos B - \sin A \sin B$,
replace $B$ by $-B$ to show that
$\cos (A - B) \equiv \cos A \cos B + \sin A \sin B$

4 Use $\dfrac{\sin (A + B)}{\cos (A + B)}$ to show that
$\tan (A + B) \equiv \dfrac{\tan A + \tan B}{1 - \tan A \tan B}$

5 Replace $B$ by $-B$ in the formula for
$\tan (A + B)$ to show that
$\tan (A - B) \equiv \dfrac{\tan A - \tan B}{1 + \tan A \tan B}$

Collecting these results we have:

$$\sin (A + B) \equiv \sin A \cos B + \cos A \sin B$$

$$\sin (A - B) \equiv \sin A \cos B - \cos A \sin B$$

$$\cos (A + B) \equiv \cos A \cos B - \sin A \sin B$$

$$\cos (A - B) \equiv \cos A \cos B + \sin A \sin B$$

$$\tan (A + B) \equiv \frac{\tan A + \tan B}{1 - \tan A \tan B}$$

$$\tan (A - B) \equiv \frac{\tan A - \tan B}{1 + \tan A \tan B}$$

### Examples 6d

1 Find the exact value of $\cos 105°$

$$\cos 105° = \cos (60° + 45°) = \cos 60° \cos 45° - \sin 60° \sin 45°$$

$$= \left(\frac{1}{2}\right)\left(\frac{\sqrt{2}}{2}\right) - \left(\frac{\sqrt{3}}{2}\right)\left(\frac{\sqrt{2}}{2}\right)$$

$$= \frac{1}{4}\left(\sqrt{2} - \sqrt{6}\right)$$

2 Prove that $\dfrac{\sin (A - B)}{\cos A \cos B} \equiv \tan A - \tan B$

Expanding the numerator, the left-hand side becomes

$$\frac{\sin A \cos B - \cos A \sin B}{\cos A \cos B}$$

$$\equiv \frac{\sin A \cos B}{\cos A \cos B} - \frac{\cos A \sin B}{\cos A \cos B}$$

$$\equiv \tan A - \tan B \equiv \text{right-hand side}$$

3 Find, in the range $0 \le \theta \le 2\pi$, the solution of the equation $2 \cos \theta = \sin \left( \theta + \frac{1}{6}\pi \right)$

$$2 \cos \theta = \sin \left( \theta + \frac{1}{6}\pi \right)$$

$$= \sin \theta \cos \frac{1}{6}\pi + \cos \theta \sin \frac{1}{6}\pi = \frac{\sqrt{3}}{2} \sin \theta + \frac{1}{2} \cos \theta$$

$$\therefore \qquad \frac{3}{2} \cos \theta = \frac{\sqrt{3}}{2} \sin \theta$$

$$\Rightarrow \qquad \frac{3}{\sqrt{3}} = \frac{\sin \theta}{\cos \theta} \qquad \Rightarrow \qquad \tan \theta = \sqrt{3}$$

Now $\quad \tan \frac{1}{3}\pi = \sqrt{3}$, so the solution is $\theta = \frac{1}{3}\pi, \frac{4}{3}\pi$

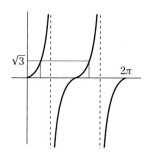

## Exercise 6d

Find the exact value of each expression, leaving your answer in surd form where necessary.

1 $\cos 40° \cos 50° - \sin 40° \sin 50°$

2 $\sin 37° \cos 7° - \cos 37° \sin 7°$

3 $\cos 75°$

4 $\tan 105°$

5 $\sin 165°$

6 $\cos 15°$

Simplify each of the following expressions.

7 $\sin \theta \cos 2\theta + \cos \theta \sin 2\theta$

8 $\cos \alpha \cos (90° - \alpha) - \sin \alpha \sin (90° - \alpha)$

9 $\dfrac{\tan A + \tan 2A}{1 - \tan A \tan 2A}$

10 $\dfrac{\tan 3\beta - \tan 2\beta}{1 + \tan 3\beta \tan 2\beta}$

Prove the following identities.

11 $\cot (A + B) \equiv \dfrac{\cot A \cot B - 1}{\cot A + \cot B}$

12 $(\sin A + \cos A)(\sin B + \cos B) \equiv \sin (A + B) + \cos (A - B)$

13 $\sin (A + B) + \sin (A - B) \equiv 2 \sin A \cos B$

14 $\cos (A + B) + \cos (A - B) \equiv 2 \cos A \cos B$

15 $\dfrac{\sin (A + B)}{\cos A \cos B} \equiv \tan A + \tan B$

Solve the following equations for values of $\theta$ in the range $0 \le \theta \le 360°$

16 $\cos (45° - \theta) = \sin \theta$

17 $3 \sin \theta = \cos (\theta + 60°)$

18 $\tan (A - \theta) = \frac{2}{3}$ and $\tan A = 3$

19 $\sin (\theta + 60°) = \cos \theta$

20 $\sin 4\theta + \sin 2\theta = 0$

## THE DOUBLE ANGLE IDENTITIES

The compound angle formulae deal with any two angles $A$ and $B$ and can therefore be used for two equal angles, i.e. when $B = A$

Replacing $B$ by $A$ in the trigonometric identities for $(A + B)$ gives the following set of double angle identities.

$$\sin 2A \equiv 2 \sin A \cos A$$

$$\cos 2A \equiv \cos^2 A - \sin^2 A$$

$$\tan 2A \equiv \frac{2 \tan A}{1 - \tan^2 A}$$

The second of these identities can be expressed in several forms because

$$\cos^2 A - \sin^2 A \equiv \begin{cases} (1 - \sin^2 A) - \sin^2 A \equiv 1 - 2 \sin^2 A \\ \cos^2 A - (1 - \cos^2 A) \equiv 2 \cos^2 A - 1 \end{cases}$$

i.e. $$\cos 2A \equiv \begin{cases} \cos^2 A - \sin^2 A \\ 1 - 2 \sin^2 A \\ 2 \cos^2 A - 1 \end{cases}$$

## Examples 6e

**1** Given $\tan \theta = \frac{3}{4}$, find the values of $\tan 2\theta$ and $\tan 4\theta$

Using $\tan 2A \equiv \dfrac{2 \tan A}{1 - \tan^2 A}$ with $A = \theta$ and $\tan \theta = \dfrac{3}{4}$ gives

$$\tan 2\theta = \frac{2\left(\frac{3}{4}\right)}{1 - \left(\frac{3}{4}\right)^2} = \frac{24}{7}$$

Using the identity for $\tan 2A$ again, but this time with $A = 2\theta$, gives

$$\tan 4\theta = \frac{2 \tan 2\theta}{1 - \tan^2 2\theta} = \frac{2\left(\frac{24}{7}\right)}{1 - \left(\frac{24}{7}\right)^2} = -\frac{336}{527}$$

**2** Eliminate $\theta$ from the equations $x = \cos 2\theta$, $y = \sec \theta$

Using $\qquad \cos 2\theta \equiv 2 \cos^2 \theta - 1$ gives

$$x = 2 \cos^2 \theta - 1 \text{ and } y = \frac{1}{\cos \theta}$$

$$x = 2\left(\frac{1}{y}\right)^2 - 1 \qquad \Rightarrow \qquad (x + 1)y^2 = 2$$

**3** Prove that $\sin 3A \equiv 3 \sin A - 4 \sin^3 A$

$$\sin 3A \equiv \sin (2A + A)$$
$$\equiv \sin 2A \cos A + \cos 2A \sin A$$
$$\equiv (2 \sin A \cos A) \cos A + (1 - 2 \sin^2 A) \sin A$$
$$\equiv 2 \sin A \cos^2 A + \sin A - 2 \sin^3 A$$
$$\equiv 2 \sin A(1 - \sin^2 A) + \sin A - 2 \sin^3 A$$
$$\equiv 3 \sin A - 4 \sin^3 A$$

**4** Find the solution of the equation $\cos 2x + 3 \sin x = 2$ giving values of $\theta$ in the interval $[-\pi, \pi]$

When a trigonometric equation involves different multiples of an angle, it is usually sensible to express the equation in a form where the trigonometric ratios are all of the same angle and, when possible, involving only one trigonometric ratio.

Using $\cos 2x \equiv 1 - 2 \sin^2 x$ gives

$$1 - 2 \sin^2 x + 3 \sin x = 2$$

$$\Rightarrow \quad 2 \sin^2 x - 3 \sin x + 1 = 0$$

$$\Rightarrow \quad (2 \sin x - 1)(\sin x - 1) = 0$$

$$\therefore \quad \sin x = \tfrac{1}{2} \quad \text{or} \quad \sin x = 1$$

When $\sin x = \tfrac{1}{2}$, $x = \tfrac{1}{6}\pi, \tfrac{5}{6}\pi$ and when $\sin x = 1$, $x = \tfrac{1}{2}\pi$

Therefore the solution is $x = \tfrac{1}{6}\pi, \tfrac{5}{6}\pi, \tfrac{1}{2}\pi$

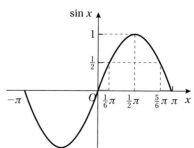

## Exercise 6e

Simplify, giving an exact value where this is possible.

**1** $2 \sin 15° \cos 15°$     **2** $\cos^2 \tfrac{1}{8}\pi - \sin^2 \tfrac{1}{8}\pi$

**3** $\sin \theta \cos \theta$     **4** $1 - 2 \sin^2 4\theta$

**5** $\dfrac{2 \tan 75°}{1 - \tan^2 75°}$     **6** $\dfrac{2 \tan 3\theta}{1 - \tan^2 3\theta}$

**7** $2 \cos^2 \tfrac{3}{8}\pi - 1$     **8** $1 - 2 \sin^2 \tfrac{1}{8}\pi$

**9** Find the value of $\cos 2\theta$ and $\sin 2\theta$ when $\theta$ is acute and when

  (a) $\cos \theta = \tfrac{3}{5}$     (b) $\sin \theta = \tfrac{7}{25}$

  (c) $\tan \theta = \tfrac{12}{5}$

**10** If $\tan \theta = -\tfrac{7}{24}$ and $\theta$ is obtuse, find

  (a) $\tan 2\theta$     (b) $\cos 2\theta$

  (c) $\sin 2\theta$     (d) $\cos 4\theta$

**11** Eliminate $\theta$ from the following pairs of equations.

  (a) $x = \tan 2\theta, y = \tan \theta$

  (b) $x = \cos 2\theta, y = \cos \theta$

  (c) $x = \cos 2\theta, y = \operatorname{cosec} \theta$

  (d) $x = \sin 2\theta, y = \sec 4\theta$

**12** Express in terms of $\cos 2x$

  (a) $2 \sin^2 x - 1$     (b) $4 - 2 \cos^2 x$

  (c) $2 \cos^2 x + \sin^2 x$

  (d) $2 \cos^2 x(1 + \cos^2 x)$

  (e) $\cos^4 x$     (Hint: $\cos^4 x \equiv (\cos^2 x)^2$)

  (f) $\sin^4 x$

**13** Prove the following identities.

  (a) $\dfrac{1 - \cos 2A}{\sin 2A} \equiv \tan A$

  (b) $\sec 2A + \tan 2A \equiv \dfrac{\cos A + \sin A}{\cos A - \sin A}$

  (c) $\cos 4A \equiv 8 \cos^4 A - 8 \cos^2 A + 1$

  (d) $\sin 2\theta \equiv \dfrac{2 \tan \theta}{1 + \tan^2 \theta}$

  (e) $\cos 2\theta \equiv \dfrac{1 - \tan^2 \theta}{1 + \tan^2 \theta}$

**14** Find the solutions of the following equations for values of $\theta$ in the range $0 \leqslant \theta \leqslant 2\pi$

  (a) $\cos 2x = \sin x$

  (b) $\sin 2x + \cos x = 0$

  (c) $\cos 2x = \cos x$

  (d) $\sin 2x = \cos x$

  (e) $4 - 5 \cos \theta = 2 \sin^2 \theta$

  (f) $\sin 2\theta - 1 = \cos 2\theta$

# f($\theta$) = $a$ cos $\theta$ + $b$ sin $\theta$

The diagrams below show the graphs of f($\theta$) = $a$ cos $\theta$ + $b$ sin $\theta$ for a variety of values of $a$ and $b$.

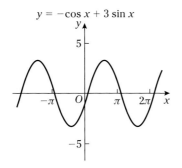

$y = -\cos x + 3\sin x$

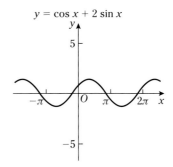

$y = \cos x + 2\sin x$

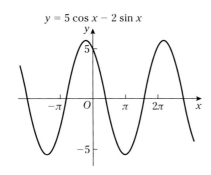

$y = 5\cos x - 2\sin x$

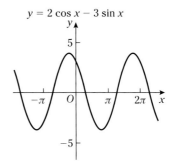

$y = 2\cos x - 3\sin x$

These diagrams suggest that it is possible to express $a$ cos $\theta$ + $b$ sin $\theta$ as $r$ sin ($\theta + \alpha$) where the values of $r$ and $\alpha$ depend on the values of $a$ and $b$. This is possible provided that we can find values of $r$ and $\alpha$ such that

$$r\sin(\theta + \alpha) \equiv a\cos\theta + b\sin\theta$$

i.e.  $\quad r\underline{\sin}\,\theta\cos\alpha + r\underline{\cos}\,\theta\sin\alpha \equiv a\underline{\underline{\cos}}\,\theta + b\underline{\sin}\,\theta$

Since this is an identity we can compare coefficients of cos $\theta$ and of sin $\theta$.

$\Rightarrow \qquad r\sin\alpha = a$ $\hspace{6cm}$ [1]

and  $\quad r\cos\alpha = b$ $\hspace{6cm}$ [2]

Equations [1] and [2] can now be solved to give $r$ and $\alpha$ in terms of $a$ and $b$.

Squaring and adding equations [1] and [2] gives

$$r^2(\sin^2\alpha + \cos^2\alpha) = a^2 + b^2 \qquad \Rightarrow \qquad r = \sqrt{a^2 + b^2}$$

Dividing equation [1] by equation [2] gives

$$\frac{r\sin\alpha}{r\cos\alpha} = \frac{a}{b} \qquad \Rightarrow \qquad \tan\alpha = \frac{a}{b}$$

Therefore $\hspace{3cm}$ **$r$ sin ($\theta + \alpha$) $\equiv$ $a$ cos $\theta$ + $b$ sin $\theta$**

$$\textbf{where } \bm{r = \sqrt{a^2 + b^2}} \textbf{ and } \bm{\tan\alpha = \frac{a}{b}}$$

It is also possible to express $a$ cos $\theta$ + $b$ sin $\theta$ as $r$ sin ($\theta - \alpha$) or as $r$ cos ($\theta \pm \alpha$), using a similar method.

## Examples 6f

1  Express $3 \sin \theta - 2 \cos \theta$ as $r \sin (\theta - \alpha)$

$$3 \sin \theta - 2 \cos \theta \equiv r \sin (\theta - \alpha)$$

$$\Rightarrow \quad 3 \sin \theta - 2 \cos \theta \equiv r \sin \theta \cos \alpha - r \cos \theta \sin \alpha$$

Comparing coefficients of $\sin \theta$ and of $\cos \theta$ gives

$$\left. \begin{array}{r} 3 = r \cos \alpha \\ 2 = r \sin \alpha \end{array} \right\} \quad \Rightarrow \quad \left\{ \begin{array}{lcl} 13 = r^2 & \Rightarrow & r = \sqrt{13} \\ \tan \alpha = \frac{2}{3} & \Rightarrow & \alpha = 33.7° \end{array} \right.$$

$$3 \sin \theta - 2 \cos \theta = \sqrt{13} \sin (\theta - 33.7°)$$

2  Find the maximum value of $f(x) = 3 \cos x + 4 \sin x$ and the smallest positive value of $x$ at which it occurs.

Expressing $f(x)$ in the form $r \sin (x + \alpha)$ enables us to 'read' its maximum value, and the values of $x$ at which it occurs, from the resulting sine wave. Note also that in this question you can choose the form in which to express $f(x)$: in this case it is sensible to choose $r \cos (x - \alpha)$ as this fits $f(x)$ better than $r \sin (x + \alpha)$ does.

$$3 \cos x + 4 \sin x \equiv r \cos (x - \alpha) \equiv r \cos x \cos \alpha + r \sin x \sin \alpha$$

Hence $\quad \left. \begin{array}{r} r \cos \alpha = 3 \\ r \sin \alpha = 4 \end{array} \right\} \quad \Rightarrow \quad \left\{ \begin{array}{lcl} r^2 = 25 & \Rightarrow & r = 5 \\ \tan \alpha = \frac{4}{3} & \Rightarrow & \alpha = 53.1° \end{array} \right.$

$$\therefore \qquad f(x) \equiv 5 \cos (x - 53.1°)$$

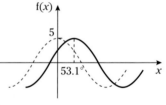

The graph of $f(x)$ is a cosine wave with a maximum value of 5 and a minimum value of $-5$ and moved $53.1°$ in the direction of the positive $x$-axis.

$\therefore$  $f(x)$ has a maximum value of 5 and, from the sketch, the smallest positive value of $x$ at which it occurs is $53.1°$.

3  Find the maximum and minimum values of $\dfrac{2}{\sin x - \cos x}$

We first express $\sin x - \cos x$ in the form $r \sin (x - \alpha)$, then the given function can be expressed as a cosec function and we can sketch its graph. Note that values of $x$ are not required so we do not need the value of $\alpha$.

When $f(x) \equiv \sin x - \cos x \equiv r \sin (x - \alpha) \equiv r \sin x \cos \alpha - r \cos x \sin \alpha$

then $\quad \left. \begin{array}{r} r \cos \alpha = 1 \\ r \sin \alpha = 1 \end{array} \right\} \quad \Rightarrow \quad r^2 = 2, \quad \text{i.e. } r = \sqrt{2}$

$\therefore \quad \sin x - \cos x \equiv \sqrt{2} \sin (x - \alpha)$

Hence $\quad \dfrac{2}{f(x)} \equiv \dfrac{2}{\sqrt{2} \sin (x - \alpha)} \equiv \sqrt{2} \operatorname{cosec} (x - \alpha)$

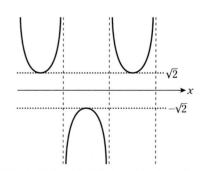

From the sketch, the maximum value of $\dfrac{2}{\sin x - \cos x}$ is $-\sqrt{2}$ and the minimum value is $\sqrt{2}$.

# THE EQUATION $a \cos x + b \sin x = c$

An equation of this type can be solved by expressing the left-hand side of the equation in the form $r \cos (x + \alpha)$ or an equivalent form. This method is illustrated in the next example.

## Examples 6f cont.

4  Find the solution of the equation for values of $x$ in the range $-\pi \leqslant x \leqslant \pi$

$$\sqrt{3} \cos x + \sin x = 1$$

If $\sqrt{3} \underline{\cos x} + \underline{\underline{\sin x}} \equiv r \cos (x - \alpha) \equiv r \underline{\cos x} \cos \alpha + r \underline{\underline{\sin x}} \sin \alpha$

then $\quad \begin{cases} r \cos \alpha = \sqrt{3} \\ r \sin \alpha = 1 \end{cases} \Rightarrow \begin{cases} r^2 = 4 \\ \tan \alpha = \dfrac{1}{\sqrt{3}} \end{cases} \Rightarrow \begin{array}{l} r = 2 \\ \alpha = \frac{1}{6}\pi \end{array}$

i.e. $\qquad \sqrt{3} \cos x + \sin x \equiv 2 \cos \left(x - \tfrac{1}{6}\pi\right)$

$\therefore \qquad$ the equation becomes

$$2 \cos \left(x - \tfrac{1}{6}\pi\right) = 1$$

$\Rightarrow \qquad \cos \left(x - \tfrac{1}{6}\pi\right) = \tfrac{1}{2}$

$\Rightarrow \qquad x - \tfrac{1}{6}\pi = \pm \tfrac{1}{3}\pi$

$\therefore \qquad x = \tfrac{1}{2}\pi, \ -\tfrac{1}{6}\pi$

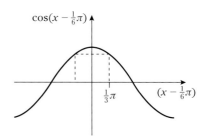

## Exercise 6f

1  Find the values of $r$ and $\alpha$ for which

(a)  $\sqrt{3} \cos \theta - \sin \theta \equiv r \cos (\theta + \alpha)$

(b)  $\cos \theta + 3 \sin \theta \equiv r \cos (\theta - \alpha)$

(c)  $4 \sin \theta - 3 \cos \theta \equiv r \sin (\theta - \alpha)$

2  Express $\cos 2\theta - \sin 2\theta$ in the form $r \cos (2\theta + \alpha)$

3  Express $2 \cos 3\theta + 5 \sin 3\theta$ in the form $r \sin (3\theta + \alpha)$

4  Express $\cos \theta - \sqrt{3} \sin \theta$ in the form $r \sin (\theta - \alpha)$. Hence sketch the graph of $f(\theta) = \cos \theta - \sqrt{3} \sin \theta$. Give the maximum and minimum values of $f(\theta)$ and the values of $\theta$ between 0 and 360° at which they occur.

5  Express $7 \cos \theta - 24 \sin \theta$ in the form $r \cos (\theta + \alpha)$. Hence sketch the graph of $f(\theta) = 7 \cos \theta - 24 \sin \theta + 3$ and give the maximum and minimum values of $f(\theta)$ and the values of $\theta$ between 0 and 360° at which they occur.

6  Find the greatest and least values of $\cos x + \sin x$. Hence find the maximum and minimum values of $\dfrac{1}{\cos x + \sin x}$

7  Find the solution of the following equations in the interval $0 \leqslant x \leqslant 360°$

(a)  $\cos x + \sin x = \sqrt{2}$

(b)  $7 \cos x + 6 \sin x = 2$

(c)  $\cos x - 3 \sin x = 1$

(d)  $2 \cos x - \sin x = 2$

## Mixed exercise 6

1   Eliminate $\theta$ from the equations
$x = \sin \theta$ and $y = \cos 2\theta$

2   Prove the identity $\dfrac{\sin 2\theta}{1 + \cos 2\theta} \equiv \tan \theta$

3   Prove that $\tan \left( \theta + \frac{1}{4} \pi \right) \tan \left( \frac{1}{4} \pi - \theta \right) \equiv 1$

4   If $\cos A = \frac{4}{5}$ and $\cos B = \frac{5}{13}$ find the possible values of $\cos (A + B)$

5   Eliminate $\theta$ from the equations
$x = \cos 2\theta$ and $y = \cos^2 \theta$

6   Prove the identity
$$\frac{\cos \theta}{1 + \sin \theta} + \frac{1 + \sin \theta}{\cos \theta} \equiv 2 \sec \theta$$

7   Express $4 \sin \theta - 3 \cos \theta$ in the form
$r \sin (\theta - \alpha)$. Hence find the maximum and minimum values of $4 \sin \theta - 3 \cos \theta + 2$

8   Express $\sin 2\theta - \cos 2\theta$ in the form
$r \sin (2\theta - \alpha)$. Hence find the smallest positive value of $\theta$ for which $\sin 2\theta - \cos 2\theta$ has a maximum value.

9   Solve the equation $\cos^2 \theta - \sin^2 \theta = 1$
for values of $\theta$ in the range $-\pi \leqslant \theta \leqslant \pi$

P3 10   Prove the identity $\cos^4 \theta - \sin^4 \theta \equiv \cos 2\theta$

P3 11   Simplify the expression $\dfrac{1 + \cos 2x}{1 - \cos 2x}$

P3 12   Find the values of $A$ between 0 and $360°$ for which $\sin (60° - A) + \sin (120° - A) = 0$

P3 13   (a)   Express $2 \sin^2 \theta + 1$ in terms of $\cos 2\theta$.

    (b)   Express $4 \cos^2 2A$ in terms of $\cos 4A$
            (Hint: use $2A = x$)

P3 14   Find all the values of $x$ between 0 and $180°$
for which $\cos x - 2 \sin x = 1$

P3 15   Solve the equation $3 \cos x - 2 \sin x = 1$ for values of $x$ in the range $0 \leqslant x \leqslant 180°$

# 7 Differentiation of trigonometric functions

*After studying this chapter you should be able to*

- use the derivatives of sin $x$, cos $x$, tan $x$, together with constant multiples, sums, differences and composites.

## INVESTIGATING THE GRADIENT FUNCTION WHEN $y = \sin x$

The diagram shows the graph of $y = \sin x$ for values of $x$ between $-\pi$ and $\frac{5}{2}\pi$. It is drawn with equal scales on the axes.

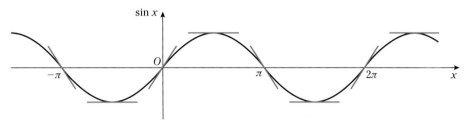

The tangents to the curve are drawn by eye and their gradients used to complete the following table.

| $x$ | $-\pi$ | $-\dfrac{\pi}{2}$ | $0$ | $\dfrac{\pi}{2}$ | $\pi$ | $\dfrac{3\pi}{2}$ | $2\pi$ | $\dfrac{5\pi}{2}$ |
|---|---|---|---|---|---|---|---|---|
| approximate gradient of sin $x$ i.e. $\dfrac{\mathrm{d}(\sin x)}{\mathrm{d}x}$ | $-1$ | $0$ | $1$ | $0$ | $-1$ | $0$ | $1$ | $0$ |

Plotting these approximate values of $\dfrac{\mathrm{d}(\sin x)}{\mathrm{d}x}$ against $x$, and drawing a smooth curve through them, gives a curve that looks like the cosine wave.

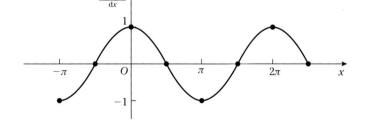

## THE DERIVATIVES OF sin $x$ AND cos $x$

When the drawing above is repeated for $y = \cos x$, where $x$ is measured in radians, and the gradient is plotted against $x$, the curve this time looks like $\dfrac{\mathrm{d}y}{\mathrm{d}x} = -\sin x$

The results are correct, i.e.

**provided $x$ is measured in radians,**

**if $y = \sin x$ then $\dfrac{\mathrm{d}y}{\mathrm{d}x} = \cos x$**

**and if $y = \cos x$ then $\dfrac{\mathrm{d}y}{\mathrm{d}x} = -\sin x$**

These two results can be quoted whenever they are needed.

## Examples 7a

**1** Find the smallest positive value of $x$ for which there is a stationary value of the function $x + 2 \cos x$

$$f(x) = x + 2 \cos x \quad \Rightarrow \quad f'(x) = 1 - 2 \sin x$$

For stationary values $\qquad f'(x) = 0$

i.e. $\qquad\qquad\qquad 1 - 2 \sin x = 0 \quad \Rightarrow \quad \sin x = \tfrac{1}{2}$

The smallest positive angle with a sine of $\tfrac{1}{2}$ is $\tfrac{1}{6}\pi$.

> The answer *must* be given in radians because the rule used to differentiate $\cos x$ is valid only for an angle in radians.

**2** Find the smallest positive value of $\theta$ for which the curve $y = 2\theta - 3 \sin \theta$ has a gradient of $\tfrac{1}{2}$

$$y = 2\theta - 3 \sin \theta \text{ gives } \frac{dy}{d\theta} = 2 - 3 \cos \theta$$

When $\qquad \dfrac{dy}{d\theta} = \tfrac{1}{2}, \qquad 2 - 3 \cos \theta = \tfrac{1}{2}$

$$3 \cos \theta = \tfrac{3}{2}$$

$$\cos \theta = \tfrac{1}{2}$$

The smallest positive value of $\theta$ for which $\cos \theta = \tfrac{1}{2}$, is $\tfrac{1}{3}\pi$.

## Exercise 7a

**1** Write down the derivative of each of the following expressions.

(a) $\sin x - \cos x$

(b) $\sin \theta + 4$

(c) $3 \cos \theta$

(d) $5 \sin \theta - 6$

(e) $2 \cos \theta + 3 \sin \theta$

(f) $4 \sin x - 5 - 6 \cos x$

**2** Find the gradient of each curve at the point whose $x$-coordinate is given.

(a) $y = \cos x; \ \tfrac{1}{2}\pi$

(b) $y = \sin x; \ 0$

(c) $y = \cos x + \sin x; \ \pi$

(d) $y = x - \sin x; \ \tfrac{1}{2}\pi$

(e) $y = 2 \sin x - x^2; \ -\pi$

(f) $y = -4 \cos x; \ \tfrac{1}{2}\pi$

**3** For each of the following curves find the smallest positive value of $\theta$ at which the gradient of the curve has the given value.

(a) $y = 2 \cos \theta; \ -1$

(b) $y = \theta + \cos \theta; \ \tfrac{1}{2}$

(c) $y = \sin \theta + \cos \theta; \ 0$

(d) $y = \sin \theta + 2\theta; \ 1$

**4** Considering only positive values of $x$, locate the first two turning points on each of the following curves and determine whether they are maximum or minimum points.

(a) $2 \sin x - x$ \qquad (b) $x + 2 \cos x$

In each case illustrate your solution by a sketch.

**5** Find the equation of the tangent to the curve $y = \cos \theta + 3 \sin \theta$ at the point where $\theta = \tfrac{1}{2}\pi$

**6** Find the equation of the normal to the curve $y = x^2 + \cos x$ at the point where $x = \pi$

**7** Find the coordinates of a point on the curve $y = \sin x + \cos x$ at which the tangent is parallel to the line $y = x$

## PRODUCTS, QUOTIENTS AND COMPOSITE FUNCTIONS

The various functions that can be differentiated when they occur in products, quotients and composite functions, now include the sine and cosine ratios.

### Differentiation of sin f(x)

When $y = \sin f(x)$ then using $u = f(x)$ gives $y = \sin u$

The chain rule gives $\quad \dfrac{dy}{dx} = \dfrac{dy}{du} \times \dfrac{du}{dx} \quad \Rightarrow \quad \dfrac{dy}{dx} = (\cos u)\dfrac{du}{dx}$

i.e. $\quad\quad\quad\quad\quad \dfrac{d}{dx} \{\sin f(x)\} = f'(x) \cos f(x)$

Similarly $\quad\quad\quad\quad \dfrac{d}{dx} \{\cos f(x)\} = -f'(x) \sin f(x)$

e.g. $\quad\quad\quad\quad \dfrac{d}{dx} \sin e^x = e^x \cos e^x \quad$ and $\quad \dfrac{d}{dx} \cos (\ln x) = -\dfrac{1}{x} \sin (\ln x)$

In particular $\quad\quad\quad \dfrac{d}{dx}(\sin ax) = a \cos ax$

and $\quad\quad\quad\quad\quad \dfrac{d}{dx}(\cos ax) = -a \sin ax$

These results are quotable.

---

### Examples 7b

**1** Differentiate $\cos \left(\frac{1}{6}\pi - 3x\right)$ with respect to $x$

$$\dfrac{d}{dx}\left\{\cos \left(\tfrac{1}{6}\pi - 3x\right)\right\} = -(-3)\sin\left(\tfrac{1}{6}\pi - 3x\right)$$

$$= 3\sin\left(\tfrac{1}{6}\pi - 3x\right)$$

**2** Find the derivative of $\dfrac{e^x}{\sin x}$

$$y = \dfrac{e^x}{\sin x} = \dfrac{u}{v} \quad \text{where} \quad u = e^x \quad \text{and} \quad v = \sin x$$

$$\Rightarrow \quad\quad \dfrac{du}{dx} = e^x \quad \text{and} \quad \dfrac{dv}{dx} = \cos x$$

$$\dfrac{dy}{dx} = \left(v\dfrac{du}{dx} - u\dfrac{dv}{dx}\right) \div v^2$$

$$= \dfrac{e^x \sin x - e^x \cos x}{\sin^2 x}$$

$$\therefore \quad \dfrac{d}{dx}\left(\dfrac{e^x}{\sin x}\right) = \dfrac{e^x}{\sin^2 x}(\sin x - \cos x)$$

**3** Find $\dfrac{dy}{d\theta}$ if $y = \cos^3 \theta$

$$y = \cos^3 \theta = [\cos \theta]^3$$

$$y = u^3 \quad \text{where} \quad u = \cos \theta$$

$$\frac{dy}{d\theta} = \frac{dy}{du} \times \frac{du}{d\theta} = (3u^2)(-\sin \theta) = 3(\cos \theta)^2(-\sin \theta)$$

$$y = \cos^3 \theta \quad \Rightarrow \quad \frac{dy}{d\theta} = -3 \cos^2 \theta \sin \theta$$

This is one example of a general rule, i.e.

$$\textbf{if } \boldsymbol{y = \cos^n x} \textbf{ then } \frac{dy}{dx} = -n \cos^{n-1} x \sin x$$

and

$$\textbf{if } \boldsymbol{y = \sin^n x} \textbf{ then } \frac{dy}{dx} = n \sin^{n-1} x \cos x$$

## Exercise 7b

Differentiate each of the following functions with respect to $x$.

1  $\sin 4x$

2  $\cos (\pi - 2x)$

3  $\sin \left(\frac{1}{2}x + \pi\right)$

4  $\dfrac{\sin x}{x}$

5  $\dfrac{\cos x}{e^x}$

6  $\sqrt{\sin x}$

7  $\sin^2 x$

8  $\sin x \cos x$

9  $e^{\sin x}$

10  $\ln (\cos x)$

11  $e^x \cos x$

12  $x^2 \sin x$

13  $\sec x$, i.e. $\dfrac{1}{\cos x}$

14  $\tan x$, i.e. $\dfrac{\sin x}{\cos x}$

15  $\csc x$

16  $\cot x$

Using the answers to questions **13** to **16**, we can now make a complete list of the derivatives of the basic trigonometric functions:

| function | derivative |
|---|---|
| $\sin x$ | $\cos x$ |
| $\cos x$ | $-\sin x$ |
| $\tan x$ | $\sec^2 x$ |
| $\cot x$ | $-\csc^2 x$ |
| $\sec x$ | $\sec x \tan x$ |
| $\csc x$ | $-\csc x \cot x$ |

## Mixed exercise 7

This exercise contains a variety of functions. Consider carefully what method to use in each case and do not forget to check first whether a given function has a standard derivative.

Find the derivative of each function in questions **1** to **12**.

**1** (a) $-\sin 4\theta$

   (b) $\theta - \cos \theta$

   (c) $\sin^3 \theta + \sin 3\theta$

**2** (a) $x^3 + e^x$

   (b) $e^{(2x+3)}$

   (c) $e^x \sin x$

**3** (a) $3 \sin x - e^{-x}$

   (b) $x^4 + 4e^x - \ln 4x$

**4** $(x+1)\ln x$

**5** $\sin^2 3x$

**6** $(4x-1)^{\frac{2}{3}}$

**7** $\dfrac{(x^4-1)}{(x+1)^3}$

**8** $x^2 \sin x$

**9** $\dfrac{e^x}{x-1}$

**10** $\dfrac{1+\sin x}{1-\sin x}$

**11** $x^2\sqrt{x-1}$

**12** $\sin x \cos^3 x$

In each question from **13** to **16**, find

(a) the gradient of the curve at the given point,

(b) the equation of the tangent to the curve at that point,

(c) the equation of the normal to the curve at that point.

**13** $y = \sin x - \cos x; \ x = \frac{1}{2}\pi$

**14** $y = x + e^x; \ x = 1$

**15** $y = 1 + x + \sin x; \ x = 0$

**16** $y = 3 - x^2 + \ln x; \ x = 1$

**17** Considering only positive values of $x$, locate the first two turning points, if there are two, on each of the following curves and determine whether they are maximum or minimum points.

   (a) $y = 1 - \sin x$     (b) $y = \frac{1}{2}x + \cos x$

**18** Find the coordinates of a point on the curve where the tangent is parallel to the given line.

   (a) $y = 3x - 2\cos x; \ y = 4x$

   (b) $y = 2\ln x - x; \ y = x$

# 8

## Implicit and parametric functions

*After studying this chapter you should be able to*

- find and use the first derivative of a function that is defined parametrically or implicitly.

## IMPLICIT FUNCTIONS

Some curves have equations that cannot easily be written as $y = f(x)$. For example it is difficult to isolate $y$ in the equation $x^2 - y^2 + y = 1$

A relationship of this type, where $y$ is not given explicitly as a function of $x$, is called an *implicit function*, because it is *implied* in the equation that $y = f(x)$

### To differentiate an implicit function

The method we use is to differentiate, term by term, with respect to $x$, but first we need to know how to differentiate terms like $y^2$ with respect to $x$.

If $\qquad g(y) = y^2 \quad$ and $\quad y = f(x)$

then $\qquad g(y) = \{f(x)\}^2$ which is a composite function.

Using the substitution $u = f(x)$ and the chain rule we have

$$\frac{d}{dx}\{u\}^2 = 2\{u\}\left(\frac{d}{dx}(u)\right)$$

Now $y = f(x) = u$ so

$$\frac{d}{dx}(y^2) = 2y\left(\frac{dy}{dx}\right) = \left(\frac{d}{dy}g(y)\right)\left(\frac{dy}{dx}\right)$$

In general, $\qquad\qquad\qquad\qquad \mathbf{\frac{d}{dx}\,g(y) = \left(\frac{d}{dy}\,g(y)\right)\left(\frac{dy}{dx}\right)}$

e.g. $\qquad \dfrac{d}{dx}y^3 = 3y^2\dfrac{dy}{dx} \quad$ and $\quad \dfrac{d}{dx}e^y = e^y\dfrac{dy}{dx}$

A term such as $xy$ is a product, so we use the product rule to differentiate such terms,

for example $\qquad \dfrac{d}{dx}(xy) = x\dfrac{d}{dx}(y) + y\dfrac{d}{dx}(x) = x\dfrac{dy}{dx} + y\,(1)$

$$= x\dfrac{dy}{dx} + y$$

We can now differentiate any expression, term by term, with respect to $x$,

e.g. if $\qquad\qquad x^2 - y^2 + x^2y = 1$

then $\qquad \dfrac{d}{dx}(x^2) - \dfrac{d}{dx}(y^2) + \dfrac{d}{dx}(x^2y) = \dfrac{d}{dx}(1)$

$\Rightarrow \qquad 2x - 2y\dfrac{dy}{dx} + 2xy + x^2\dfrac{dy}{dx} = 0$

Hence $\qquad 2x(1 + y) = \dfrac{dy}{dx}(2y - x^2) \qquad \Rightarrow \qquad \dfrac{dy}{dx} = \dfrac{2x(1 + y)}{2y - x^2}$

## Examples 8a

1  Differentiate each equation with respect to $x$ and hence find $\dfrac{dy}{dx}$ in terms of $x$ and $y$

(a)  $x^3 + xy^2 - y^3 = 5$

(b)  $y = xe^y$

(a)  When $x^3 + xy^2 - y^3 = 5$ differentiating term by term gives

$$\frac{d}{dx}(x^3) + \frac{d}{dx}(xy^2) - \frac{d}{dx}(y^3) = \frac{d}{dx}(5)$$

$$\therefore \quad 3x^2 + y^2 + 2xy\frac{dy}{dx} - 3y^2\frac{dy}{dx} = 0$$

Hence $\quad \dfrac{dy}{dx} = \dfrac{(3x^2 + y^2)}{y(3y - 2x)}$

(b)  When $y = xe^y, \dfrac{dy}{dx} = \dfrac{d}{dx}(xe^y)$

$$= e^y\frac{d}{dx}(x) + x\frac{d}{dx}(e^y)$$

$$\Rightarrow \quad \frac{dy}{dx} = e^y + xe^y\frac{dy}{dx}$$

Hence $\quad \dfrac{dy}{dx} = \dfrac{e^y}{1 - xe^y}$

2  Find the equation of the tangent to the curve $x^2 + xy - y^2 = 4$ at the point $(2, 4)$ on the curve.

$$x^2 + xy - y^2 = 4 \quad \Rightarrow \quad 2x + x\frac{dy}{dx} + y - 2y\frac{dy}{dx} = 0$$

When $x = 2$ and $y = 4$, $4 + 2\dfrac{dy}{dx} + 4 - 8\dfrac{dy}{dx} = 0 \quad \Rightarrow \quad \dfrac{dy}{dx} = \dfrac{4}{3}$

We need first to find the gradient of the curve, i.e. $\dfrac{dy}{dx}$, at the point $(2, 4)$.

The equation of the tangent is given by $y - 4 = \frac{4}{3}(x - 2)$

$$\Rightarrow \quad 4x - 3y + 4 = 0$$

## Exercise 8a

Differentiate the following equations with respect to $x$.

1  $x^2 + y^2 = 4$

2  $x^2 + xy + y^2 = 0$

3  $x(x + y) = y^2$

4  $\dfrac{1}{x} + \dfrac{1}{y} = e^y$

5  $\sin x + \sin y = 1$

6  $xe^y = x + 1$

7  Find $\dfrac{dy}{dx}$ as a function of $x$ if $y^2 = 2x + 1$

8  Find the gradient of $x^2 + y^2 = 9$ at the points where $x = 1$

9  Find the equation of the tangent at $\left(1, \frac{1}{3}\right)$ to the curve whose equation is

$$2x^2 + 3y^2 - 3x + 2y = 0$$

10  The equation of a curve is $\cos x + \sin y = 1$

(a)  Find $\dfrac{dy}{dx}$ in terms of $x$ and $y$.

(b)  At the point $\left(\frac{\pi}{4}, \alpha\right)$ on the curve, the gradient of the tangent is 1. Find the value of $\alpha$ for $0 \leqslant \alpha \leqslant \pi$

**P3 11**   The equation of a curve is $x^3 + y^3 = 28$

   (a)   Find $\dfrac{dy}{dx}$ in terms of $x$ and $y$.

   (b)   Find the value of $y$ when $x = 1$ and hence find the equation of the tangent to the curve at this point.

**P3 12**   The equation of a curve is $x^2 + y^2 = \alpha^2$

   (a)   Find $\dfrac{dy}{dx}$ in terms of $x$ and $y$.

   (b)   Find the equation of the tangent to the curve at the point $\left( \alpha\sqrt{2}, \alpha\sqrt{2} \right)$.

**P3 13**   The equation of a curve is $x^2 + 2y^2 = 3$

Find the equation of the normal to the curve at the point on the curve at which $x = y$

## PARAMETRIC EQUATIONS

When a direct relationship between $x$ and $y$ is awkward to work with, it may be easier to express $x$ and $y$ each in terms of a third variable, called a *parameter*.

For example, the equations

$$x = t^2, y = t - 1$$

A point $P(x, y)$ is on the curve given by these equations if and only if the coordinates of $P$ are $(t^2, t - 1)$.

By giving $t$ any value we choose, we get a pair of corresponding values of $x$ and $y$. For example, when $t = 3$, $x = 9$ and $y = 2$, therefore $(9, 2)$ is a point on the curve. By giving $t$ several other values, we can plot points and draw the curve.

The direct relationship between $x$ and $y$ can be found by eliminating $t$ from these two *parametric equations*. In this case it is $(y + 1)^2 = x$

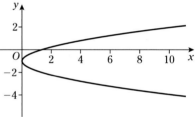

### Finding the gradient function using parametric equations

When $x = \mathrm{f}(t)$ and $y = \mathrm{g}(t)$, we can use the chain rule, $\dfrac{dy}{dx} = \dfrac{dy}{dt} \times \dfrac{dt}{dx}$ together with $\dfrac{dt}{dx} = \dfrac{1}{\frac{dx}{dt}}$ to give

$$\mathbf{\dfrac{dy}{dx} = \dfrac{\dfrac{dy}{dt}}{\dfrac{dx}{dt}}}$$

Hence, for the parametric equations $x = t^2$ and $y = t - 1$, we have

$$\dfrac{dy}{dt} = 1, \dfrac{dx}{dt} = 2t \quad \Rightarrow \quad \dfrac{dy}{dx} = \dfrac{1}{2t}$$

Each point on the curve is defined by a value of $t$ that also gives the value of $\dfrac{dy}{dx}$ at that point.

For example, when $t = 1$, $\dfrac{dy}{dx} = \dfrac{1}{2}$.

Conversely, the value(s) of $t$ where $\dfrac{dy}{dx}$ has a particular value lead to the coordinates of the relevant point(s) on the curve.

For example, when $\dfrac{dy}{dx} = 2$, $\dfrac{1}{2t} = 2 \quad \Rightarrow \quad t = \frac{1}{4}$ and when $t = \frac{1}{4}$, $x = \frac{1}{16}$ and $y = -\frac{3}{4}$;

i.e. the gradient of the curve is 2 at the point $\left( \frac{1}{16}, -\frac{3}{4} \right)$.

There are no values of $t$ for which $\dfrac{dy}{dx} = 0$, so there are no stationary points on this curve.

## Examples 8b

1  Find the stationary point on the curve whose parametric equations are $x = t^3$, $y = (t + 1)^2$ and determine its nature.

$$\frac{dy}{dx} = \frac{\dfrac{dy}{dt}}{\dfrac{dx}{dt}} = \frac{2(t + 1)}{3t^2}$$

At stationary points $\dfrac{dy}{dx} = 0$    i.e. $t = -1$

When   $t = -1$, $x = -1$   and   $y = 0$

Therefore the stationary point is $(-1, 0)$.

| Value of $x$ | $-2$ | $-1$ | $-\frac{1}{2}$ |
|---|---|---|---|
| Value of $t$ | $-\sqrt[3]{2}$ | $-1$ | $-\dfrac{1}{\sqrt[3]{2}}$ |
| Sign of $\dfrac{dy}{dx}$ | $-$ | $0$ | $+$ |

To determine the nature of the stationary point we examine the sign of $\dfrac{dy}{dx}$ near the point by first choosing appropriate values for $x$ and then finding the corresponding values of $t$.

Therefore $(-1, 0)$ is a minimum point.   $\dfrac{d^2y}{dx^2}$ is awkward to work with when $x$ and $y$ are given in terms of a parameter so it is not sensible to use it as a test for the nature of a stationary point.

2  The parametric equations of a curve are $x = 2t$, $y = t^2$
   Find the equation of the tangent to the curve at the point $(2t, t^2)$.

$$x = 2t,\ y = t^2 \quad \text{gives} \quad \frac{dy}{dx} = \frac{2t}{2} = t$$

Using   $y - y_1 = m(x - x_1)$  gives  $y - t^2 = t(x - 2t)$

i.e.     $y = tx - t^2$

By finding the equation of the tangent to the curve in terms of $t$, we can find the equation of the tangent at any given value of $t$. For example, when $t = 2$, the equation of the tangent is $y = 2x - 4$

## Exercise 8b

1  Find the gradient function of each of the following curves in terms of the parameter.

   (a)  $x = 2t^2$, $y = t$

   (b)  $x = \cos\theta$, $y = \sin\theta$

   (c)  $x = t$, $y = \dfrac{4}{t}$

2  If $x = \dfrac{t}{1 - t}$ and $y = \dfrac{t^2}{1 - t}$, find $\dfrac{dy}{dx}$ in terms of $t$.

   What is the value of $\dfrac{dy}{dx}$ at the point where $x = 1$?

3  If $x = t^2$ and $y = t^3$, find $\dfrac{dy}{dx}$ in terms of $t$.

4  The parametric equations of a curve are $y = e^t - 1$ and $x = 1 - e^{-t}$

   Find $\dfrac{dy}{dx}$ in terms of $t$.

5  Find the turning points of the curve whose parametric equations are $x = t$, $y = t^3 - t$, and distinguish between them.

6  A curve has parametric equations

   $$x = \theta - \cos\theta,\ y = \sin\theta.$$

   Find the smallest positive value of $\theta$ at which the gradient of this curve is zero.

7  Find the equation of the tangent to the curve $x = t^2$, $y = 4t$ at the point where $t = -1$

8  Find the equation of the normal to the curve

   $$x = 2\cos\theta,\ y = 3\sin\theta$$

   at the point where $\theta = \frac{1}{4}\pi$.
   Find the coordinates of the point where this normal cuts the curve again.

## Mixed exercise 8

1 Differentiate with respect to $x$

   (a) $y^4$       (b) $xy^2$

   (c) $\dfrac{1}{y}$       (d) $x \ln y$

   (e) $\sin y$       (f) $e^y$

   (g) $y \cos x$       (h) $y \cos y$

In questions **2** to **4** find $\dfrac{dy}{dx}$ in terms of $x$ and $y$.

2 $x^2 - 2y^2 = 4$

3 $\dfrac{1}{x} + \dfrac{1}{y} = 2$

4 $x^2 y^3 = 9$

In questions **5** to **10** find $\dfrac{dy}{dx}$ in terms of the parameter.

5 $x = t^2, y = t^3$

6 $x = (t + 1)^2, y = t^2 - 1$

7 $x = \sin^2 \theta, y = \cos^3 \theta$

8 $x = 4t, y = \dfrac{4}{t}$

9 $x = e^t, y = 1 - t$

10 $x = \dfrac{t}{1 - t}, y = \dfrac{t^2}{1 - t}$

11 If $x = \sin t$ and $y = \cos 2t$, find $\dfrac{dy}{dx}$ in terms of $x$.

12 If $x = e^t - t$ and $y = e^{2t} - 2t$, show that
$$\frac{dy}{dx} = 2(e^t + 1)$$

13 Differentiate $y^2 - 2xy + 3y = 7x$ with respect to $x$. Hence find the equation of the tangent to the curve at the point where $y = 1$

14 Find the equation of the normal to the curve
$$x = \cos \theta, y = 2 \sin \theta$$
at the point where $\theta = \dfrac{3\pi}{4}$

# 9 Integration 1

## After studying this chapter you should be able to

- extend the idea of 'reverse differentiation' to include the integration of $e^{ax+b}$, $\dfrac{1}{ax+b}$, $\sin(ax+b)$, $\cos(ax+b)$ and $\sec^2(ax+b)$
- use trigonometrical relationships (such as double-angle formulae) to facilitate the integration of functions such as $\cos^2 x$
- use the trapezium rule to estimate the value of a definite integral, and use sketch graphs in simple cases to determine whether the trapezium rule gives an overestimate or an underestimate.

### Differentiation reversed

We know from Pure Mathematics 1, Chapter 12 (page 111) that

$$\int (ax+b)^n \, dx = \frac{1}{(a)(n+1)}(ax+b)^{n+1} + K$$

We now look at more functions whose integrals can be found by recognising them as the differentials of known functions.

## INTEGRATING EXPONENTIAL FUNCTIONS

It is already known that $\dfrac{d}{dx} e^x = e^x$

$\therefore$
$$\int e^x dx = e^x + K$$

Also $\qquad \dfrac{d}{dx}(ce^x) = ce^x$

and $\qquad \dfrac{d}{dx} e^{(ax+b)} = ae^{(ax+b)}$

Hence $\quad \displaystyle\int ce^x dx = ce^x + K \qquad$ and $\qquad \displaystyle\int e^{(ax+b)} = \frac{1}{a}e^{(ax+b)} + K$

e.g. $\quad \displaystyle\int 2e^x dx = 2e^x + K \qquad$ and $\qquad \displaystyle\int 4e^{(1-3x)}dx = (4)\left(-\frac{1}{3}\right)e^{(1-3x)} + K$

---

### Example 9a

Write down the integral of $e^{3x}$ with respect to $x$ and hence evaluate $\displaystyle\int_0^1 e^{3x}\,dx$

$$\int e^{3x}\,dx = \tfrac{1}{3}e^{3x} + K$$

The constant of integration disappears when a definite integral is calculated, hence

$$\int_0^1 e^{3x}dx = \left[\tfrac{1}{3}e^{3x}\right]_0^1 = \tfrac{1}{3}e^3 - \tfrac{1}{3}e^0$$

i.e. $\quad \displaystyle\int_0^1 e^{3x}\,dx = \tfrac{1}{3}(e^3 - 1)$

## Exercise 9a

Integrate each function with respect to $x$.

**1** $e^{4x}$

**2** $4e^{-x}$

**3** $e^{(3x-2)}$

**4** $2e^{(1-5x)}$

**5** $6e^{-2x}$

**6** $5e^{(x-3)}$

**7** $e^{\left(2+\frac{x}{2}\right)}$

**8** $e^{2x} + \dfrac{1}{e^{2x}}$

Evaluate the following definite integrals.

**9** $\displaystyle\int_0^2 e^{2x}dx$

**10** $\displaystyle\int_{-1}^1 2e^{(x+1)}dx$

**11** $\displaystyle\int_2^3 e^{(2-x)}dx$

**12** $\displaystyle\int_0^2 -e^xdx$

# TO INTEGRATE $\dfrac{1}{x}$

We might think that we can write $\dfrac{1}{x} = x^{-1}$ and integrate by using the rule $\displaystyle\int x^n\,dx = \dfrac{1}{n+1}x^{(n+1)} + K$, but this method fails when $n = -1$ because the resulting integral is meaningless.

Taking a second look at $\dfrac{1}{x}$, it is *recognised* as the derivative of $\ln x$.

However, $\ln x$ is defined only when $x > 0$. Hence, provided that $x > 0$ we have

$$\dfrac{d}{dx}(\ln x) = \dfrac{1}{x} \quad \Leftrightarrow \quad \int \dfrac{1}{x}\,dx = \ln x + K$$

When $x < 0$ the statement $\displaystyle\int \dfrac{1}{x}\,dx = \ln x$ is not true because the log of a negative number does not exist.

However, the function $\dfrac{1}{x}$ exists for negative values of $x$, as the graph of $y = \dfrac{1}{x}$ shows.

Also, the definite integral $\displaystyle\int_c^d \dfrac{1}{x}\,dx$, which is represented by the shaded area, clearly exists. It must, therefore, be possible to integrate $\dfrac{1}{x}$ when $x$ is negative.

If $x < 0$ then $-x > 0$

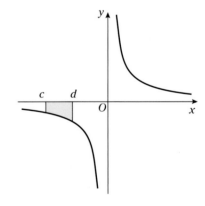

i.e.  $\displaystyle\int \dfrac{1}{x}\,dx = \int \dfrac{-1}{(-x)}\,dx = \ln(-x) + K$

Therefore when $x < 0$, $\displaystyle\int \dfrac{1}{x}\,dx = \ln(-x) + K$

and when $x > 0$, $\displaystyle\int \dfrac{1}{x}\,dx = \ln x + K$

These two results can be combined using $|x|$ so that, for both positive and negative values of $x$, we have

$$\int \dfrac{1}{x}\,dx = \ln|x| + K$$

We also know that $\dfrac{d}{dx}\ln(ax + b) = \dfrac{a}{ax + b}$

$\therefore \quad \displaystyle\int \dfrac{1}{ax+b}\,dx = \dfrac{1}{a}\ln|ax+b| + K$  e.g.  $\displaystyle\int \dfrac{1}{2x+5}\,dx = \tfrac{1}{2}\ln|2x+5| + K$

and  $\displaystyle\int \dfrac{1}{4-3x}\,dx = -\tfrac{1}{3}\ln|4-3x| + K$

## Exercise 9b

Integrate with respect to $x$

1  $\dfrac{1}{2x}$

2  $\dfrac{2}{x}$

3  $\dfrac{1}{4x}$

4  $\dfrac{3}{2x}$

5  $\dfrac{4}{x-1}$

6  $\dfrac{1}{3x+1}$

7  $\dfrac{3}{1-2x}$

8  $\dfrac{6}{2+3x}$

9  $\dfrac{3}{4-2x}$

10 $\dfrac{4}{1-x}$

11 $\dfrac{5}{6-7x}$

Evaluate

12 $\displaystyle\int_{1}^{2} \dfrac{3}{x+1}\,\mathrm{d}x$

13 $\displaystyle\int_{1}^{2} \dfrac{1}{2x-1}\,\mathrm{d}x$

14 $\displaystyle\int_{4}^{5} \dfrac{2}{x-3}\,\mathrm{d}x$

15 $\displaystyle\int_{0}^{1} \dfrac{1}{2-x}\,\mathrm{d}x$

## INTEGRATING TRIGONOMETRIC FUNCTIONS

Knowing the derivatives of the six trigonometric functions, we can recognise the following integrals.

$$\dfrac{\mathrm{d}}{\mathrm{d}x}(\sin x) = \cos x \qquad \Leftrightarrow \qquad \int \cos x\,\mathrm{d}x = \sin x + K$$

$$\dfrac{\mathrm{d}}{\mathrm{d}x}(\cos x) = -\sin x \qquad \Leftrightarrow \qquad \int \sin x\,\mathrm{d}x = -\cos x + K$$

$$\dfrac{\mathrm{d}}{\mathrm{d}x}(\tan x) = \sec^2 x \qquad \Leftrightarrow \qquad \int \sec^2 x\,\mathrm{d}x = \tan x + K$$

Remembering the derivatives of some variations of the basic trigonometric functions, we also have

$$\int c \cos x\,\mathrm{d}x = c \sin x + K$$

and  $$\int \cos (ax+b)\,\mathrm{d}x = \dfrac{1}{a}\sin (ax+b) + K$$

with similar results for the remaining trigonometric integrals

e.g.  $$\int 3 \sec^2 x\,\mathrm{d}x = 3 \tan x + K$$

$$\int \sin 4\theta\,\mathrm{d}\theta = -\tfrac{1}{4}\cos 4\theta + K$$

You do not need to learn these, you should recognise them as the differentials of standard trigonometric functions.

## Exercise 9c

Integrate each function with respect to $x$.

1  $\sin 2x$

2  $\cos 7x$

3  $\sec^2 4x$

4  $\sin\left(\frac{1}{4}\pi + x\right)$

5  $3\cos\left(4x - \frac{1}{2}\pi\right)$

6  $\sec^2\left(\frac{1}{3}\pi + 2x\right)$

7  $3\sin 5x$

8  $2\sin(3x - \alpha)$

9  $5\cos\left(\alpha - \frac{1}{2}x\right)$

10  $\sin(4x - \pi)$

11  $\cos 3x - \cos x$

12  $\sec^2 2x$

Evaluate

13  $\displaystyle\int_0^{\frac{\pi}{6}} \sin 3x \, dx$

14  $\displaystyle\int_{\frac{\pi}{4}}^{\frac{\pi}{6}} \cos\left(2x - \frac{1}{2}\pi\right) dx$

15  $\displaystyle\int_0^{\frac{\pi}{2}} 2\sin\left(2x - \frac{1}{2}\pi\right) dx$

16  $\displaystyle\int_0^{\frac{\pi}{8}} \sec^2 2x \, dx$

17  Differentiate $\ln\cos x$ with respect to $x$.

Hence find $\displaystyle\int_0^{\frac{\pi}{3}} \tan x \, dx$

## Integration of powers of sin $x$, cos $x$ and tan $x$

When an integral of a trigonometric function is not one of the standard integrals, using a trigonometric identity often helps to change the integral to one that we can recognise.

The double angle formulae $\cos 2x = 1 - 2\sin^2 x$ and $\cos 2x = 2\cos^2 x - 1$ are useful when finding integrals of powers of $\sin x$ and $\cos x$.

---

### Examples 9d

1  (a)  Show that $4 - 2\cos^2 x = 5 - \cos 2x$

(b)  Hence find $\displaystyle\int(4 - 2\cos^2 x)\, dx$

(a)  $4 - 2\cos^2 x = 4 - (\cos 2x - 1) = 5 - \cos 2x$      Using $\cos 2A = 2\cos^2 A - 1$

(b)  $\displaystyle\int(4 - 2\cos^2 x)\, dx = \int(5 - \cos 2x)\, dx = 5x + \frac{1}{2}\sin 2x + K$

2  (a)  Differentiate $\cos^3 x$ with respect to $x$.

(b)  Hence find $\displaystyle\int \sin x \cos^2 x \, dx$

(a)  Using $y = \cos^3 x$, the chain rule with $u = \cos x$ gives

$$\frac{dy}{dx} = \frac{dy}{du} \times \frac{du}{dx} = 3u^2 \times -\sin x = -3\cos^2 x \sin x$$

(b)  $\displaystyle\int \sin x \cos^2 x \, dx$

$\qquad = -\frac{1}{3}\cos^3 x + K$

## Exercise 9d

1  Given that $3 \sin x - 4 \sin^3 x = \sin 3x$, find
$\int (3 \sin x - 4 \sin^3 x)\, dx$

2  Use the identity $\dfrac{2 \tan x}{1 + \tan^2 x} \equiv \sin 2x$

to find the exact value of $\displaystyle\int_0^{\frac{\pi}{4}} \dfrac{2 \tan x}{1 + \tan^2 x}\, dx$

3  Use the identity $8 \cos^4 x - 8 \cos^2 x + 1 \equiv \cos 4x$
to find $\int (8 \cos^4 x - 8 \cos^2 x + 1)\, dx$

4  (a)  Express $2 \sin^2 x + 1$ in terms of $\cos 2x$

   (b)  Use the result of (a) to find $\int \sin^2 x\, dx$

5  Use the double angle formulae for $\cos 2x$ and
$\sin 2x$ to show that
$2 \cos^2 x + 4 \sin x \cos x = \cos 2x + 2 \sin 2x - 1$
Hence find $\int (2 \cos^2 x + 4 \sin x \cos x)\, dx$

6  Express $\tan^2 x$ in terms of $\sec^2 x$

Hence find the exact value of $\displaystyle\int_{\frac{\pi}{4}}^{0} 4 \tan^2 x\, dx$

7  Differentiate $\sin^3 x$ with respect to $x$.
Hence find $\int (6 \cos x \sin^2 x)\, dx$

## THE TRAPEZIUM RULE

We know that the definite integral $\displaystyle\int_a^b f(x)\, dx$ can be used to evaluate the area between the curve

$y = f(x)$, the $x$-axis and the ordinates at $x = a$ and $x = b$. But we cannot always find a function whose derivative is $f(x)$. In such cases the definite integral, and hence the exact value of the specified area, cannot be found.

However, if we divide the area into a *finite* number of strips, then the sum of their areas gives an approximate value for the required area and hence an approximate value of the definite integral.

When the area shown in the diagram is divided into vertical strips, each strip is approximately a trapezium.

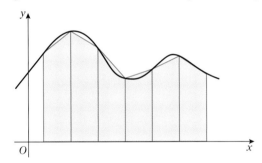

When the width of the strip and its two vertical sides are known, the area of the trapezium-shaped strip can be found using the formula

   area $= \frac{1}{2}$ (sum of parallel sides) $\times$ width

The sum of the areas of all the strips then gives an approximate value for the area under the curve.

When there are $n$ strips, *all with the same width*, $d$ say, and the vertical edges of the strips
(i.e. the ordinates) are labelled $y_0, y_1, y_2, \ldots, y_{n-1}, y_n$

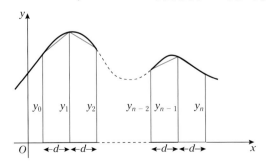

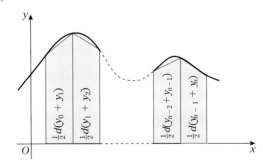

then the sum of the areas of all the strips is

$$\tfrac{1}{2}(y_0 + y_1)(d) + \tfrac{1}{2}(y_1 + y_2)(d) + \tfrac{1}{2}(y_2 + y_3)(d) + \dots + \tfrac{1}{2}(y_{n-2} + y_{n-1})(d) + \tfrac{1}{2}(y_{n-1} + y_n)(d)$$

Therefore the area, $A$, under the curve is given approximately by

$$A \approx \tfrac{1}{2}(d)\,[y_0 + 2y_1 + 2y_2 + \dots 2y_{n-1} + y_n]$$

This formula is known as the *trapezium rule*.

An easy way to remember the formula in terms of ordinates is

**half width of strip × (first + last + twice all the others)**

The width of a strip is an interval along the $x$-axis.

Be careful: the number of intervals and the number of ordinates are not the same.

---

### Example 9e

Use the trapezium rule with four intervals to find an approximate value for the definite integral $\displaystyle\int_1^5 x^3\,dx$.

Is the result an overestimate or an underestimate of $\displaystyle\int_1^5 x^3\,dx$ ?

The given definite integral represents the area bounded by the $x$-axis, the lines $x = 1$ and $x = 5$, and the curve $y = x^3$

Five ordinates are used when there are four strips whose widths must all be the same. From $x = 1$ to $x = 5$ there are four units so the width of each strip must be 1 unit. Hence the five ordinates are where $x = 1, x = 2, x = 3, x = 4$ and $x = 5$

Using the trapezium rule,

$$y_0 = 1^3 = 1, \quad y_1 = 2^3 = 8, \quad y_2 = 3^3 = 27, \quad y_3 = 64, \quad y_4 = 125$$

The required area, $A$, is given by

$$A \approx \tfrac{1}{2}(1)[1 + 125 + 2\{8 + 27 + 64\}] = 162$$

The required area is approximately 162 square units.

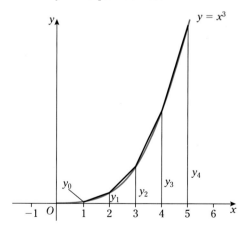

From the sketch, the area of each trapezium is greater than the area under that part of the curve, therefore 162 is an overestimate of $\displaystyle\int_1^5 x^3\,dx$.

## Exercise 9e

Use the trapezium rule with two intervals to find an approximate value for each definite integral.

**1** $\displaystyle\int_1^3 \frac{1}{x^2}\,dx$

**2** $\displaystyle\int_1^3 \ln x\,dx$

**3** $\displaystyle\int_0^{\frac{2\pi}{3}} \sqrt{\sin x}\,dx$

**4** $\displaystyle\int_1^3 xe^{-x}\,dx$

**5** Use sketch graphs to determine whether the results for questions **1** and **2** are underestimates or overestimates.

**6** The diagram shows a sketch of the curve
$$y = \frac{2x}{\ln x}$$

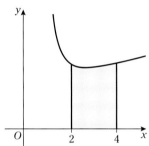

(a) Use the trapezium rule with two intervals to find an approximate value of the shaded area.

(b) Explain whether your approximation is an underestimate or an overestimate of the shaded area.

**7** The diagram shows a sketch of part of the curve $y = x\cos x$

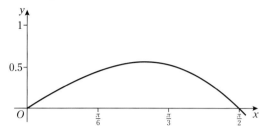

(a) Use the trapezium rule with three intervals to find an approximate value of
$$\int_0^{\frac{\pi}{2}} \cos x\,dx$$

(b) Use the sketch to determine whether your result is an underestimate or overestimate.

## Mixed exercise 9

Integrate the following functions with respect to $x$.

**1** $3e^{(2x-1)}$

**2** $\dfrac{1}{3x}$

**3** $\cos(2x + \pi)$

**4** $\dfrac{2}{1+x}$

**5** $\sec^2(2x - 1)$

**6** $e^x + e^{-x}$

**7** $\sin\left(\dfrac{\pi}{3} - 3x\right)$

**8** $\dfrac{4}{3x-2}$

Evaluate

**9** $\displaystyle\int_0^1 3e^{4x-1}\,dx$

**10** $\displaystyle\int_0^{\frac{\pi}{8}} \cos 4x\,dx$

**11** $\displaystyle\int_3^4 \frac{1}{x-2}\,dx$

**12** $\displaystyle\int_{\frac{\pi}{4}}^{\frac{\pi}{2}} \sin\left(2x - \frac{\pi}{4}\right)dx$

**13** Use trigonometrical identities to evaluate
$$\int_0^{\frac{\pi}{4}} (\cos^2 x + 2\sin^2 x)\,dx$$

**P3 14** (a) Use the expansions of $\sin(4x + x)$ and $\sin(4x - x)$ to show that
$$2\sin 4x \cos x \equiv \sin 5x + \sin 3x$$

(b) Use the result of part (a) or otherwise to find the exact value of
$$\int_0^{\frac{\pi}{2}} \sin 4x \cos x\,dx$$

# 10 Numerical solutions of equations

## After studying this chapter you should be able to

- locate approximately a root of an equation, by means of graphical considerations and/or searching for a sign change
- understand the idea of, and use the notation for, a sequence of approximations which converges to a root of an equation
- understand how a given simple iterative formula of the form $x_{n+1} = f(x_n)$ relates to the equation being solved, and use a given iteration, or an iteration based on a given rearrangement of an equation, to determine a root to a prescribed degree of accuracy.

## APPROXIMATE SOLUTIONS

When the roots of an equation cannot be found exactly, we can look for approximate solutions. First we need to locate the roots roughly and this can be done by sketching graphs.

For example, the roots of the equation $e^x = 4x$ are the values of $x$ where the curve $y = e^x$ and the line $y = 4x$ intersect.

From the sketch we can see that there is one root between 0 and 0.5 and another root somewhere near 2.

We now need a way to locate the roots more accurately.

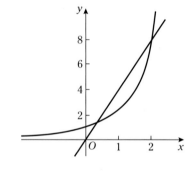

## LOCATING THE ROOTS OF AN EQUATION

The roots of the equation $f(x) = 0$ are the values of $x$ where the curve $y = f(x)$ crosses the $x$-axis, e.g.

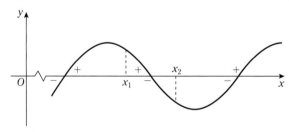

Each time the curve crosses the $x$-axis, the sign of $y$ changes. So

> **if one root only of the equation $f(x) = 0$ lies between $x_1$ and $x_2$, and if the curve $y = f(x)$ is unbroken between the points where $x = x_1$ and $x = x_2$, then $f(x_1)$ and $f(x_2)$ are opposite in sign.**

The condition that the curve $y = f(x)$ must be unbroken between $x_1$ and $x_2$ is essential, as we can see from the curve on the right.

This curve crosses the $x$-axis between $x_1$ and $x_2$ but $f(x_1)$ and $f(x_2)$ have the same sign because the curve is broken between these values.

Returning to the equation $e^x = 4x$, we will now locate the larger root a little more precisely.

First we write the equation in the form $f(x) = 0$, i.e. $e^x - 4x = 0$, then we find where there is a change in the sign of $f(x)$.

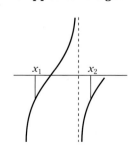

We know that there is a root in the region of $x = 2$, so we will see whether it lies between 1.8 and 2.2.

Using    $f(x) = e^x - 4x$    gives    $f(1.8) = e^{1.8} - 4(1.8) = -1.1...$

and    $f(2.2) = e^{2.2} - 4(2.2) = 0.2...$

Therefore the larger root of the equation lies between 1.8 and 2.2 (and is likely to be nearer to 2.2 as $f(2.2)$ is nearer to zero than $f(1.8)$ is).

---

## Example 10a

Find the turning points on the curve $y = x^3 - 2x^2 + x + 1$. Hence sketch the curve and use the sketch to show that the equation $x^3 - 2x^2 + x + 1 = 0$ has only one real root. Find two consecutive integers between which this root lies.

At turning points $\dfrac{dy}{dx} = 0$,   i.e. $3x^2 - 4x + 1 = 0$

$\Rightarrow$    $(3x - 1)(x - 1) = 0$ so $x = \frac{1}{3}$ and 1

When $x = \frac{1}{3}$, $y = \frac{31}{27}$ and when $x = 1$, $y = 1$

As the curve is a cubic, and as $y \to \infty$ as $x \to \infty$, we deduce that the curve has a maximum point at $\left(\frac{1}{3}, \frac{31}{27}\right)$ and a minimum point at $(1, 1)$.

From the sketch, we see that the curve $y = x^3 - 2x^2 + x + 1$ crosses the $x$-axis at one point only, therefore the equation $x^3 - 2x^2 + x + 1 = 0$ has only one real root.

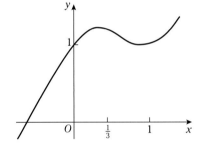

Also from the sketch it appears that this root lies between $x = -1$ and $x = 0$

Using $f(x) = x^3 - 2x^2 + x + 1$ gives $f(-1) = -3$ and $f(0) = 1$

As $f(-1)$ and $f(0)$ are opposite in sign, the one real root of $x^3 - 2x^2 + x + 1 = 0$ lies between $x = -1$ and $x = 0$

---

## Exercise 10a

1 Use sketch graphs to determine the number of real roots of each equation. (Some may have an infinite set of roots.)

(a) $\sin x = \dfrac{1}{x}$

(b) $\cos x = x^2 - 1$

(c) $x^2 = \tan x$

(d) $\sin x = x$

(e) $(x^2 - 4) = \dfrac{1}{x}$

(f) $\sin x = x^2$

(g) $x \ln x = 1$

(h) $xe^x = 1$

(i) $\ln x + e^x = 0$

2 For each equation in question 1 with a finite number of roots, locate the root, or the larger root where there is more than one, within an interval of half a unit.

3 Find the turning points on the curve whose equation is $y = x^3 - 3x^2 + 1$. Hence sketch the curve and use your sketch to find the number of real roots of the equation $x^3 - 3x^2 + 1 = 0$

4 Using a method similar to that given in question **3**, or otherwise, determine the number of real roots of each equation.

(a) $x^4 - 3x^3 + 1 = 0$

(b) $x^3 - 24x + 1 = 0$

(c) $x^5 - 5x^2 + 4 = 0$

5 Show that the equation $e^{-x} = x^2 + 2$ has just one root and find this root to the nearest integer.

6 Find the successive integers between which the smallest root of the equation $e^x = \frac{1}{2}(x + 3)$ lies.

## ITERATION

Iteration produces a sequence of values by using a formula (called an iteration formula) of the form $x_{n+1} = f(x_n)$

Taking $x_1$ as the first value, then

$$x_2 = f(x_1)$$
$$x_3 = f(x_2)$$
$$x_4 = f(x_3)$$
$$x_5 = f(x_4) \quad \text{and so on.}$$

For example, when

$$x_{n+1} = (x_n + 1)^{\frac{1}{2}} \quad \text{and} \quad x_1 = 2$$
$$x_2 = (2 + 1)^{\frac{1}{2}} = 1.732...$$
$$x_3 = (1.732... + 1)^{\frac{1}{2}} = 1.652...$$
$$x_4 = (1.652... + 1)^{\frac{1}{2}} = 1.628...$$
$$x_5 = (1.628... + 1)^{\frac{1}{2}} = 1.621...$$

and so on.

This sequence of values converges to a value $\alpha$, because as $n$ increases, $x_n$ gets closer and closer to $x_{n+1}$, i.e. $x_n \to \alpha$. This value, $\alpha$, is when $x_n = x_{n+1}$ i.e. when $\alpha = (\alpha + 1)^{\frac{1}{2}}$. Therefore $\alpha$ is a root of the equation $\alpha = (\alpha + 1)^{\frac{1}{2}}$

Not all iterations give values that converge.

For example, using the iteration formula, $x_{n+1} = (x_n + 1)^2$, and taking $x_1 = 1$

$$x_2 = (1 + 1)^2 = 4$$
$$x_3 = (4 + 1)^2 = 25$$
$$x_4 = (25 + 1)^2 = 676$$

This sequence of values diverges because the values are increasing (rapidly in this case).

### Using iteration to find an approximate value for a root of an equation

We can often use an iteration formula to find a good approximation to a root, $\alpha$, of an equation $f(x) = 0$

When $f(x) = 0$ can be written in the form $x = F(x)$ we use this to make the iteration formula $x_{n+1} = F(x_n)$

For example, we will look at the equation $e^{x+1} - x - 3 = 0$

The graph of $y = e^{x+1} - x - 3$ shows that the equation has two roots, one, $\alpha$, near $-3$ and the other, $\beta$, near $0$.

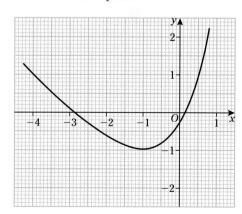

Rearranging $e^{x+1} - x - 3 = 0$ as $x = e^{x+1} - 3$ and changing this to the iteration formula $x_{n+1} = e^{x_n+1} - 3$ we can start with $x_1 = -3$ to give the sequence

$$x_2 = e^{-2} - 3 = -2.8646...$$
$$x_3 = e^{-1.8646...} - 3 = -2.9439...$$
$$x_4 = e^{-1.8450...} - 3 = -2.7472...$$
$$x_5 = e^{-1.8519...} - 3 = -2.9475...$$

We can now see that this sequence is converging to a value that, correct to 2 decimal places, is $-2.84$.

Therefore $\alpha = -2.95$ correct to 2 d.p.

We can continue the iteration if we need a more accurate value for $\alpha$.

Using the same iteration formula with $x_1 = 0$ gives the sequence

$$x_2 = e^1 - 3 = -0.2817...$$
$$x_3 = e^{1.0497...} - 3 = -0.9490...$$
$$x_4 = e^{0.8575...} - 3 = -1.9477...$$
$$x_5 = e^{0.3561...} - 3 = -2.6123...$$

This time the sequence is diverging away from zero, so it fails to find the root near 0.

The diagrams show how this iteration works.

This diagram shows the graphs of $y = x$ and $y = e^{x+1} - 3$, and shows the points of intersection near $x = -3$ and $x = 0$

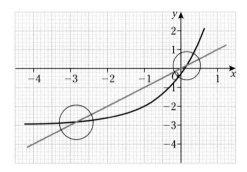

These values of $x$ are where $x = e^{x+1} - 3$

i.e. they are the roots of the equation $e^{x+1} - x - 3 = 0$

These two graphs show enlargements near the two points of intersection.

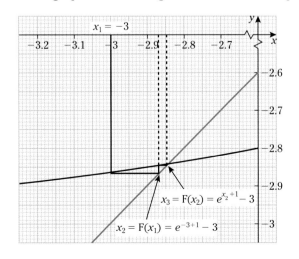

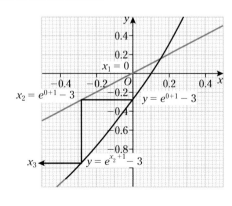

### Examples 10b

**1** (a) Show by calculation that the equation $\cos x - 2x = 0$ has a root between $x = 0$ and $x = \frac{\pi}{4}$

(b) Show that the sequence of values, if they converge, given by the iterative formula
$x_{n+1} = \frac{1}{2} \cos x_n$ converges to the value of the root of the equation in part (a).

(c) Starting with $x_1 = 0.5$, find this root correct to 2 decimal places.

(a) When $x = 0$, $\cos x - 2x = 1$

When $x = \frac{\pi}{4}$, $\cos x - 2x = \cos \frac{\pi}{4} - \frac{\pi}{2} = -0.86\ldots$

$f(0)$ and $f\left(\frac{\pi}{4}\right)$ have opposite signs, therefore there is a root between $x = 0$ and $x = \frac{\pi}{4}$

(b) $x_{n+1} = \frac{1}{2} \cos x_n \Rightarrow \cos x_n - 2x_{n+1} = 0$ and if $x_n \to \alpha$, then $\cos \alpha - 2\alpha = 0$,
i.e. $\alpha$ is a root of the equation.

(c) To give an answer correct to 2 d.p., give the intermediate values to 4 d.p.

$x_1 = 0.5$ gives $x_2 = \frac{1}{2} \cos 0.5 = 0.4387\ldots$

$x_3 = \frac{1}{2} \cos 0.4387 = 0.4526\ldots$

$x_4 = \frac{1}{2} \cos 0.4526 = 0.4496\ldots$

$x_5 = \frac{1}{2} \cos 0.4496 = 0.4502\ldots$

$x_6 = \frac{1}{2} \cos 0.4502 = 0.4501\ldots$

$x_3$ and $x_4$ both give 0.45 correct to 2 d.p., so we could at this stage check that the root lies between 0.449 and 0.450.

Therefore the root is 0.45 correct to 2 decimal places.

**2** The equation $x^3 - 2x + 1 = 0$ has a root between $-2$ and $-1$. Use the iteration formula
$x_{n+1} = \frac{1}{2}(x_n^3 + 1)$ to show that, starting with $x_1 = -1.5$, the iteration fails.

$x_2 = \frac{1}{2}(-1.5^3 + 1) = -1.1875$

$x_3 = \frac{1}{2}(-1.1875^3 + 1) = -0.337\ldots$ This is outside the interval in which the root lies.

Therefore the iteration fails.

**3** The sequence of values given by the iteration formula $x_{n+1} = \frac{1}{5}(1 - 2x_n^2 - x_n^3)$ with $x_1 = 0$ converges to a value $\alpha$.

(a) Find $\alpha$ correct to 3 decimal places.

(b) Give the equation, one of whose roots is $\alpha$.

(a) $x_2 = \frac{1}{5}(1 - 2(0)^2 - (0)^3) = 0.2$

$x_3 = \frac{1}{5}(1 - 2(0.2)^2 - (0.2)^3) = 0.1824$

$x_4 = \frac{1}{5}(1 - 2(0.1824)^2 - (0.1824)^3) = 0.185478\ldots$

$x_2 = \frac{1}{5}(1 - 2(0.185478)^2 - (0.185478)^3) = 0.184962\ldots$

$x_6 = \frac{1}{5}(1 - 2(0.184962)^2 - (0.184962)^3) = 0.185049\ldots$

$\alpha = 0.185$ to 3 d.p.

(b) $\alpha$ is a root of the equation $x = \frac{1}{5}(1 - 2x^2 - x^3)$ which simplifies to $x^3 + 2x^2 + 5x - 1 = 0$

## Exercise 10b

Show by calculation that each of the following equations has a root between 0 and 1.

**1**  $x^3 - x^2 + 10x - 2 = 0$

**2**  $3x^3 - 2x^2 - 9x + 2 = 0$

**3**  $2x^3 + x^2 + 6x - 1 = 0$

**4**  $x^2 + 8x - 8 = 0$

**5**  The sequence of values given by the iteration formula $x_n = \sqrt[3]{3x_n + 3}$ converges to $\alpha$.

Starting with $x_1 = 2$, find $x_2, x_3$ and $x_4$ giving your answers to 4 decimal places.

**6**  (a)  For the equation in question **1**, explain why $x_{n+1} = \frac{1}{10}(2 + x_n^2 - x_n^3)$ is an iteration formula that gives a sequence that converges to the root between 0 and 1.

(b)  Use the formula with $x_1 = 0.5$ to find this root correct to 2 decimal places.

**7**  (a)  Show that equation $x \ln x = 1$ has a root between 1.5 and 2.

(b)  Show that the sequence given by the iteration formula $x_{n+1} = \dfrac{1}{\ln x_n}$ with $x_1 = 1.5$ does not converge.

**8**  The sequence of values given by the iteration formula $x_{n+1} = \dfrac{4}{3} + \dfrac{4}{3x_n}$ converges to $\alpha$ when $x_1 = 1.8$

(a)  Find the value of $\alpha$ correct to 1 decimal place.

(b)  Write down the equation for which $x = \alpha$ is a solution. Hence find the exact value of $\alpha$.

**9**  Starting with $x_1 = 3$, the sequence of values given by the iteration formula $x_{n+1} = \ln x_n + 2$ converges to $\alpha$.

(a)  Find the value of $\alpha$ correct to 2 decimal places.

(b)  Write down the equation for which $x = \alpha$ is a solution.

**P3 10**  Use an iteration formula to find the root of the equation given in question **3** correct to 2 decimal places.

**P3 11**  Find iteration formulae for questions **1**, **2** and **4**. Determine in each case whether the iteration converges or fails.

# Summary 2

## TRIGONOMETRY

### Compound angle identities

$$\sin (A \pm B) \equiv \sin A \cos B \pm \cos A \sin B$$
$$\cos (A \pm B) \equiv \cos A \cos B \mp \sin A \sin B$$
$$\tan (A \pm B) \equiv \frac{\tan A \pm \tan B}{1 \mp \tan A \tan B}$$

### Double angle identities

$$\sin 2A \equiv 2 \sin A \cos A$$

$$\cos 2A \equiv \begin{cases} \cos^2 A - \sin^2 A \\ 2 \cos^2 A - 1 \\ 1 - 2 \sin^2 A \end{cases}$$

$$\text{and} \quad \begin{cases} \cos^2 A \equiv \frac{1}{2}(1 + \cos 2A) \\ \sin^2 A \equiv \frac{1}{2}(1 - \cos 2A) \end{cases}$$

$$\tan 2A \equiv \frac{2 \tan A}{1 - \tan^2 A}$$

### Expressing $a \cos \theta \pm b \sin \theta$ as a single term

For various values of $a$ and $b$, $a \cos \theta \pm b \sin \theta$ can be expressed as

$$r \cos (\theta \pm \alpha) \quad \text{or} \quad r \sin (\theta \pm \alpha)$$

where $r = \sqrt{a^2 + b^2}$ and $\tan \alpha$ is either $\dfrac{a}{b}$ or $\dfrac{b}{a}$

## DIFFERENTIATION

### Standard results

| $f(x)$ | $\frac{d}{dx} f(x)$ |
|---|---|
| $\sin x$ | $\cos x$ |
| $\cos x$ | $-\sin x$ |
| $\tan x$ | $\sec^2 x$ |
| $\sin (ax + b)$ | $a \cos (ax + b)$ |
| $\cos (ax + b)$ | $-a \sin (ax + b)$ |

## IMPLICIT DIFFERENTIATION

When $y$ cannot be isolated, each term can be differentiated with respect to $x$. Remember that

$$\frac{d}{dx} (y) = \frac{dy}{dx} \text{ and that } \frac{d}{dx} [f(y)] = f'(y) \frac{dy}{dx}$$

e.g. $\dfrac{d}{dx} (y^2) = (2y) \left( \dfrac{dy}{dx} \right)$

and $\dfrac{d}{dx} (xy) = y + (x) \left( \dfrac{dy}{dx} \right)$ (by product rule)

## Parametric differentiation

If $y = f(t)$ and $x = g(t)$ then $\dfrac{dy}{dx} = \dfrac{dy}{dt} \div \dfrac{dx}{dt}$

## Standard integrals

| Function | Integral |
|---|---|
| $e^{ax+b}$ | $\dfrac{1}{a} e^{ax+b}$ |
| $\dfrac{a}{x}, \quad \dfrac{1}{ax+b}$ | $a \ln |x|, \quad \dfrac{1}{a} \ln |ax+b|$ |
| $\cos x$ | $\sin x$ |
| $\sin x$ | $-\cos x$ |
| $\sec^2 x$ | $\tan x$ |

## The trapezium rule

$$A \approx \int_a^b f(x)\, dx \approx \tfrac{1}{2} d[y_0 + 2y_1 + \ldots + 2y_{n-1} + y_n]$$

where the values of $y$ are the lengths of the parallel sides of the trapeziums, i.e. the $y$-coordinates of the points on the curve at the edge of each strip.

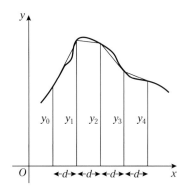

## Locating the root of an equation

An approximate solution can be found from a graph.

To show that $f(x) = 0$ has a root between $x_1$ and $x_2$ use the fact that $f(x) = 0$ between two points on the $x$-axis where $f(x_1) < 0$ and $f(x_2) > 0$ provided that $f(x)$ has no breaks between $x_1$ and $x_2$

## Iteration

Rearrange $f(x) = 0$ as $x = g(x)$ then change this to the iteration formula $x_{n+1} = g(x_n)$

## Summary exercise 2

1  The sequence of values given by the iterative formula

$$x_{n+1} = \frac{1}{5}\left(4x_n + \frac{306}{x_n^4}\right)$$

with initial value $x_1 = 3$ converges to $\alpha$.

(i) Use this iterative formula to find $\alpha$ correct to 3 decimal places, showing the result of each iteration. [3]

(ii) State an equation satisfied by $\alpha$, and hence show that the exact value of $\alpha$ is $\sqrt[5]{306}$. [2]

Cambridge, Paper 2 Q2 J04

2  (i) Express $3 \sin \theta + 4 \cos \theta$ in the form $R \sin (\theta + \alpha)$, where $R > 0$ and $0° < \alpha \leqslant 90°$, giving the value of $\alpha$ correct to 2 decimal places. [3]

(ii) Hence solve the equation

$$3 \sin \theta + 4 \cos \theta = 4.5$$

giving all solutions in the interval $0° \leqslant \theta \leqslant 360°$, correct to 1 decimal place. [4]

(iii) Write down the least value of $3 \sin \theta + 4 \cos \theta + 7$ as $\theta$ varies. [1]

Cambridge, Paper 2 Q4 J04

**3** The parametric equations of a curve are

$$x = 3t + \ln(t - 1), \quad y = t^2 + 1, \quad \text{for } t > 1$$

(i) Express $\dfrac{dy}{dx}$ in terms of $t$. [3]

(ii) Find the coordinates of the only point on the curve at which the gradient of the curve is equal to 1. [4]

Cambridge, Paper 2 Q3 J07

**4** The equation of a curve is

$$x^2 + y^2 - 4xy + 3 = 0$$

(i) Show that $\dfrac{dy}{dx} = \dfrac{2y - x}{y - 2x}$ [4]

(ii) Find the coordinates of each of the points on the curve where the tangent is parallel to the $x$-axis. [5]

Cambridge, Paper 2 Q7 J08

**5**

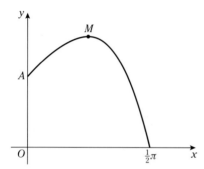

The diagram shows the part of the curve $y = e^x \cos x$ for $0 \le x \le \frac{1}{2}\pi$. The curve meets the $y$-axis at the point $A$. The point $M$ is a maximum point.

(i) Write down the coordinates of $A$. [1]

(ii) Find the $x$-coordinate of $M$. [4]

(iii) Use the trapezium rule with three intervals to estimate the value of

$$\int_0^{\frac{1}{2}\pi} e^x \cos x \, dx,$$

giving your answer correct to 2 decimal places. [3]

(iv) State, with a reason, whether the trapezium rule gives an underestimate or an overestimate of the true value of the integral in part (iii). [1]

Cambridge, Paper 2 Q7 J07

**6** (i) Express $5\cos\theta - \sin\theta$ in the form $R\cos(\theta + \alpha)$, where $R > 0$ and $0° < \alpha < 90°$, giving the exact value of $R$ and the value of $\alpha$ correct to 2 decimal places. [3]

(ii) Hence solve the equation

$$5\cos\theta - \sin\theta = 4$$

giving all solutions in the interval $0° \le \theta \le 360°$ [4]

Cambridge, Paper 2 Q5 J08

**7** Solve the equation $\sec x = 4 - 2\tan^2 x$, giving all solutions in the interval $0° \le x \le 180°$ [6]

Cambridge, Paper 2 Q5 J09

**8** Show that $\displaystyle\int_0^6 \dfrac{1}{x + 2} \, dx = 2\ln 2$ [4]

Cambridge, Paper 21 Q2 J10

**9**

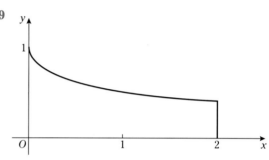

The diagram shows the curve $y = \dfrac{1}{1 + \sqrt{x}}$ for values of $x$ from 0 to 2.

(i) Use the trapezium rule with two intervals to estimate the value of

$$\int_0^2 \dfrac{1}{1 + \sqrt{x}} \, dx,$$

giving your answer correct to 2 decimal places. [3]

(ii) State, with a reason, whether the trapezium rule gives an underestimate or an overestimate of the true value of the integral in part (i). [1]

Cambridge, Paper 2 Q3 J00

**10** (i) By sketching a suitable pair of graphs, show that the equation

$$e^{2x} = 2 - x$$

has only one root. [3]

(ii) Verify by calculation that this root lies between $x = 0$ and $x = 0.5$ [2]

(iii) Show that, if a sequence of values given by the iterative formula

$$x_{n+1} = \tfrac{1}{2}\ln(2 - x_n)$$

converges, then it converges to the root of the equation in part (i). [1]

(iv) Use this iterative formula, with initial value $x_1 = 0.25$, to determine the root correct to 2 decimal places. Give the result of each iteration to 4 decimal places.   [3]

*Cambridge, Paper 21 Q7 J10*

**11** The angle $x$, measured in degrees, satisfies the equation

$$\cos (x - 30) = 3 \sin (x - 60°)$$

(i) By expanding each side, show that the equation may be simplified to

$$(2\sqrt{3}) \cos x = \sin x \qquad [3]$$

(ii) Find the two possible values of $x$ lying between $0°$ and $360°$.   [3]

(iii) Find the exact value of $\cos 2x$, giving your answer as a fraction.   [3]

*Cambridge, Paper 2 Q5 N02*

**12** The equation of a curve is

$$2x^2 + 3y^2 - 2xy = 10$$

(i) Show that $\dfrac{dy}{dx} = \dfrac{y - 2x}{3y - x}$   [4]

(ii) Find the coordinates of the points on the curve where the tangent is parallel to the $x$-axis.   [5]

*Cambridge, Paper 2 Q7 N02*

**13** (i) By differentiating $\dfrac{\cos x}{\sin x}$, show that if $y = \cot x$ then $\dfrac{dy}{dx} = -\operatorname{cosec}^2 x$   [3]

(ii) Hence show that $\displaystyle\int_{\frac{1}{6}\pi}^{\frac{1}{2}\pi} \operatorname{cosec}^2 x \, dx = \sqrt{3}$   [2]

By using appropriate trigonometrical identities, find the exact value of

(iii) $\displaystyle\int_{\frac{1}{6}\pi}^{\frac{1}{2}\pi} \cot^2 x \, dx$   [3]

(iv) $\displaystyle\int_{\frac{1}{6}\pi}^{\frac{1}{2}\pi} \dfrac{1}{1 - \cos 2x} \, dx$   [3]

*Cambridge, Paper 2 Q7 N03*

**14**

The diagram shows the curve $y = x^2 e^{-x}$ and its maximum point $M$.

(i) Find the $x$-coordinate of $M$.   [4]

(ii) Show that the tangent to the curve at the point where $x = 1$, passes through the origin.   [3]

(iii) Use the trapezium rule, with two intervals, to estimate the value of

$$\int_1^3 x^2 e^{-x} \, dx,$$

giving your answer correct to 2 decimal places.   [3]

*Cambridge, Paper 2 Q8 N07*

**15** (i) Prove the identity

$$(\cos x + 3 \sin x)^2 \equiv 5 - 4 \cos 2x + 3 \sin 2x \qquad [4]$$

(ii) Using the identity, or otherwise, find the exact value of

$$\int_0^{\frac{1}{4}\pi} (\cos x + 3 \sin x)^2 \, dx.$$   [4]

*Cambridge, Paper 2 Q7 N07*

**16** The equation of a curve is $y^2 + 2xy - x^2 = 2$

(i) Find the coordinates of the two points on the curve where $x = 1$   [2]

(ii) Show by differentiation that at one of these points the tangent to the curve is parallel to the $x$-axis. Find the equation of the tangent to the curve at the other point, giving your answer in the form $ax + by + c = 0$   [7]

*Cambridge, Paper 2 Q8 N09*

**17** (i) Show that the equation

$$\sin (x + 30°) = 2 \cos (x + 60°)$$

can be written in the form

$$(3\sqrt{3}) \sin x = \cos x \qquad [3]$$

(ii) Hence solve the equation

$$\sin (x + 30°) = 2 \cos (x + 60°)$$

for $-180° < x < 180°$   [3]

*Cambridge, Paper 21 Q4 N08*

**18** The curve with equation $y = x \ln x$ has one stationary point.

(i) Find the exact coordinates of this point, giving your answers in terms of e.   [5]

(ii) Determine whether this point is a maximum or a minimum point.   [2]

*Cambridge, Paper 2 Q6 N09*

# 11 Partial fractions and the binomial theorem

*After studying this chapter you should be able to*

- recall an appropriate form for expressing rational functions in partial fractions, and carry out the decomposition, in cases where the denominator is no more complicated than

  $(ax + b)(cx + d)(ex + f)$,
  $(ax + b)(cx + d)^2$,
  $(ax + b)(x^2 + c^2)$

  and where the degree of the numerator does not exceed that of the denominator
- use the expansion of $(1 + x)^n$, where $n$ is a rational number and $|x| < 1$

## PARTIAL FRACTIONS

We can express two separate fractions as a single fraction with a common denominator. Now we are going to reverse this process, i.e. we will take an expression such as $\dfrac{x - 2}{(x + 3)(x - 4)}$ and express it as the sum of two separate fractions.

This process is called decomposing into *partial fractions*.

The fraction $\dfrac{x - 2}{(x + 3)(x - 4)}$

is a *proper fraction* because the highest power of $x$ in the numerator (1 in this case) is less than the highest power of $x$ in the denominator (2 in this case when the brackets are expanded).

Therefore its separate (or partial) fractions also will be proper,

i.e. $\dfrac{x - 2}{(x + 3)(x - 4)}$ can be expressed as $\dfrac{A}{x + 3} + \dfrac{B}{x - 4}$

where $A$ and $B$ are numbers. The worked example which follows shows how the values of $A$ and $B$ can be found.

### Examples 11a

1 Express $\dfrac{x - 2}{(x + 3)(x - 4)}$ in partial fractions.

$$\frac{x - 2}{(x + 3)(x - 4)} \equiv \frac{A}{x + 3} + \frac{B}{x - 4}$$

Express the separate fractions on the right-hand side as a single fraction over a common denominator.

$$\frac{x - 2}{(x + 3)(x - 4)} \equiv \frac{A(x - 4) + B(x + 3)}{(x + 3)(x - 4)}$$

This is an identity because the right-hand side is just another way of expressing the left-hand side. It follows that, as the denominators are identical, the numerators also are identical.

$$\Rightarrow \quad x - 2 \equiv A(x - 4) + B(x + 3)$$

Remembering that this is two ways of writing the same expression, it follows that the sides are equal for any value that we choose to give to $x$.

Choosing to substitute 4 for $x$ (to eliminate $A$) gives

$$2 = A(0) + B(7) \quad \Rightarrow \quad B = \tfrac{2}{7}$$

Choosing to substitute $-3$ for $x$ (to eliminate $B$) gives

$$-5 = A(-7) + B(0)$$

$$\Rightarrow \quad A = \tfrac{5}{7}$$

Therefore $\quad \dfrac{x - 2}{(x + 3)(x - 4)} \equiv \dfrac{\tfrac{5}{7}}{x + 3} + \dfrac{\tfrac{2}{7}}{x - 4}$

$$\equiv \dfrac{5}{7(x + 3)} + \dfrac{2}{7(x - 4)}$$

This method can be extended to decomposing a fraction with more than two linear factors in the denominator into partial fractions.

**2** Express $\dfrac{3x^2 + 19x - 32}{(x - 1)(x - 2)(x + 4)}$ in partial fractions.

First express the fraction as three separate fractions, then express the separate fractions as a single fraction over a common denominator.

$$\dfrac{3x^2 + 19x - 32}{(x - 1)(x - 2)(x + 4)} = \dfrac{A}{x - 1} + \dfrac{B}{x - 2} + \dfrac{C}{x + 4}$$

$$= \dfrac{A(x - 2)(x + 4) + B(x - 1)(x + 4) + C(x - 1)(x - 2)}{(x - 1)(x - 2)\,(x + 4)}$$

The numerators of the single fractions are identical,

$$\therefore \quad 3x^2 + 19x - 32 = A(x - 2)(x + 4) + B(x - 1)(x + 4) + C(x - 1)(x - 2)$$

Choose values of $x$ that eliminate each of the constants in turn.

$x = 2$ gives $\qquad 18 = B(1)(6) \Rightarrow B = 3$

$x = -4$ gives $\qquad -60 = C(-5)(-6) \Rightarrow C = -2$

$x = 1$ gives $\qquad -10 = A(-1)(5) \Rightarrow A = 2$

Therefore

$$\dfrac{3x^2 + 19x - 32}{(x - 1)(x - 2)(x + 4)} = \dfrac{2}{x - 1} + \dfrac{3}{x - 2} - \dfrac{2}{x + 4}$$

## Exercise 11a

Express the following fractions in partial fractions.

**1** $\dfrac{x - 2}{(x + 1)(x - 1)}$

**2** $\dfrac{2x - 1}{(x - 1)(x - 7)}$

**3** $\dfrac{4}{(x + 3)(x - 2)}$

**4** $\dfrac{7x}{(2x - 1)(x + 4)}$

**5** $\dfrac{2}{x(x - 2)}$

**6** $\dfrac{2x - 1}{x^2 - 3x + 2}$

**7** $\dfrac{3}{x^2 - 9}$

**8** $\dfrac{6x + 7}{3x(x + 1)}$

**9** $\dfrac{9}{2x^2 + x}$

**10** $\dfrac{x + 1}{3x^2 - x - 2}$

**11** $\dfrac{2}{(x + 1)(x - 1)}$

**12** $\dfrac{3}{(x - 2)(x + 1)}$

**13** $\dfrac{1}{x(x - 3)}$

**14** $\dfrac{4}{(x - 1)(x + 3)}$

**15** $\dfrac{1}{(x^2 - 1)}$

**16** $\dfrac{2}{(2x + 1)(2x - 1)}$

**17** $\dfrac{3x^2 + 4x - 1}{(x - 1)(x + 1)(x + 2)}$

**18** $\dfrac{x}{(x - 2)(x + 2)(x - 1)}$

## Quadratic factors in the denominator

We can also decompose fractions with quadratic factors in the denominator.

For example   $\dfrac{x^2 + 1}{(x^2 + 2)(x - 1)}$

is a proper fraction, so its partial fractions are also proper, i.e.

$\dfrac{x^2 + 1}{(x^2 + 2)(x - 1)}$   can be expressed in the form   $\dfrac{Ax + B}{x^2 + 2} + \dfrac{C}{x - 1}$

Expressing this as a single fraction and comparing the numerators gives

$$x^2 + 1 \equiv (Ax + B)(x - 1) + C(x^2 + 2)$$

When $x = 1$,  $2 = C(3)$   $\Rightarrow$   $C = \frac{2}{3}$

The values of $A$ and $B$ can then be found by substituting any suitable values for $x$.

We will choose $x = 0$ and $x = -1$ as these are simple values to handle. (We do not choose $x = 1$ as it was used to find $C$.)

$x = 0$ gives   $1 = B(-1) + \frac{2}{3}(2)$   $\Rightarrow$   $B = \frac{1}{3}$

$x = -1$ gives   $2 = \left(-A + \frac{1}{3}\right)(-2) + \frac{2}{3}(3)$   $\Rightarrow$   $A = \frac{1}{3}$

$\therefore$   $\dfrac{x^2 + 1}{(x^2 + 2)(x - 1)} \equiv \dfrac{x + 1}{3(x^2 + 2)} + \dfrac{2}{3(x - 1)}$

## A repeated factor in the denominator

The fraction $\dfrac{2x - 1}{(x - 2)^2}$ is a proper fraction, and can be expressed as two fractions with numerical numerators, as we can see if we adjust the numerator,

i.e.   $\dfrac{2x - 1}{(x - 2)^2} \equiv \dfrac{2(x - 2) - 1 + 4}{(x - 2)^2} \equiv \dfrac{2}{x - 2} + \dfrac{3}{(x - 2)^2}$

Any fraction whose denominator is a repeated linear factor can be expressed as separate fractions with numerical numerators.

When the numerator is not so easy to rearrange, the values of the numerators can be found using the method in the next worked example.

To summarise, a proper fraction can be decomposed into partial fractions and the form of the partial fractions depends on the form of the factors in the denominator where

a linear factor gives a partial fraction of the form $\dfrac{A}{ax + b}$

a quadratic factor gives a partial fraction of the form $\dfrac{Ax + B}{ax^2 + bx + c}$

a repeated factor gives two partial fractions of the form $\dfrac{A}{ax + b} + \dfrac{B}{(ax + b)^2}$

---

### Example 11b

Express $\dfrac{x - 1}{(x + 1)(x - 2)^2}$ in partial fractions.

$$\frac{x - 1}{(x + 1)(x - 2)^2} \equiv \frac{A}{(x + 1)} + \frac{B}{(x - 2)} + \frac{C}{(x - 2)^2}$$

$$\Rightarrow \quad x - 1 \equiv A(x - 2)^2 + B(x + 1)(x - 2) + C(x + 1)$$

$$x = -1 \text{ gives } A = -\tfrac{2}{9}$$

$$x = 2 \text{ gives } C = \tfrac{1}{3}$$

Comparing coefficients of $x^2$ gives $0 = -\tfrac{2}{9} + B \quad \Rightarrow \quad B = \tfrac{2}{9}$

$$\therefore \quad \frac{x - 1}{(x + 1)(x - 2)^2} \equiv -\frac{2}{9(x + 1)} + \frac{2}{9(x - 2)} + \frac{1}{3(x - 2)^2}$$

---

### Exercise 11b

Express in partial fractions

1  $\dfrac{2}{(x - 1)(x + 1)^2}$

2  $\dfrac{x^2 + 3}{x(x^2 + 2)}$

3  $\dfrac{2x^2 + x + 1}{(x - 3)(x + 1)^2}$

4  $\dfrac{x^2 + 1}{x(2x^2 + 1)}$

5  $\dfrac{x}{(x - 1)(x - 2)^2}$

6  $\dfrac{(x^2 - 1)}{x^2(2x + 1)}$

7  $\dfrac{x^2 - 2}{(x + 3)(x - 1)^2}$

8  $\dfrac{(x - 1)}{(x + 1)(x + 2)^2}$

9  Express $\dfrac{x}{(x^2 - 4)(x - 1)}$ in partial fractions

    (a)  by first treating $(x^2 - 4)$ as a quadratic factor

    (b)  by first factorising $(x^2 - 4)$.

State which method you think is better and explain why.

## Improper fractions

When the highest power in the numerator is greater than or equal to the highest power in the denominator, the fraction is improper.

For example, $\dfrac{x^2 - 2}{(x + 3)(x - 2)}$ is an improper fraction.

Before we decompose this fraction into partial fractions, we turn it into a constant plus a proper fraction. We do this by adjusting the numerator so that part of it is equal to the denominator, i.e.

$$\frac{x^2 - 2}{(x + 3)(x - 2)} = \frac{x^2 - 2}{x^2 + x + 6} = \frac{(x^2 + x - 6) - x + 4}{x^2 + x - 6}$$

We add $x$ and subtract 4 to change $x^2 - 2$ to $x^2 + x - 6$, then subtract $x$ and add 4 to keep the value of the numerator the same.

$$= 1 + \frac{-x + 4}{x^2 + x - 6}$$

We can then decompose $\dfrac{-x + 4}{x^2 + x - 6}$ into partial fractions.

---

### Example 11c

Express $\dfrac{x^2}{(x + 1)(x - 3)}$ in partial fractions.

This fraction is improper and it must be rearranged to obtain a mixed number before it can be expressed in partial fractions.

$$\frac{x^2}{(x + 1)(x - 3)} = \frac{x^2}{x^2 - 2x - 3} = \frac{x^2 - 2x - 3 + 2x + 3}{x^2 - 2x - 3}$$

$$= 1 + \frac{2x + 3}{(x + 1)(x - 3)}$$

$$\frac{2x + 3}{(x + 1)(x - 3)} \equiv \frac{A}{x + 1} + \frac{B}{x - 3} \quad \Rightarrow \quad 2x + 3 \equiv A(x - 3) + B(x + 1)$$

$x = 3$ gives $B = \frac{9}{4}$, $x = -1$ gives $A = -\frac{1}{4}$

$$\therefore \qquad \frac{x^2}{(x + 1)(x - 3)} = 1 - \frac{1}{4(x + 1)} + \frac{9}{4(x - 3)}$$

---

### Exercise 11c

Express in partial fractions.

**1** $\dfrac{x^2}{(x + 1)(x - 1)}$

**2** $\dfrac{x^2 + 3}{(x - 1)(x + 1)}$

**3** $\dfrac{x^2 - 2}{(x + 3)(x - 1)}$

**4** $\dfrac{x^3}{(x + 2)(x^2 + 1)}$

---

### Mixed exercise 11d

**1** Express the given fraction in partial fractions.

(a) $\dfrac{4}{(2x + 1)(x - 3)}$

(b) $\dfrac{(3x - 2)}{(x + 1)(4x - 3)}$

(c) $\dfrac{2t}{(t^2 - 1)}$

**2** Express in partial fractions

(a) $\dfrac{3}{x(2x + 1)}$

(b) $\dfrac{x + 4}{(x + 3)(x - 5)}$

(c) $\dfrac{(2x - 3)}{(x - 2)(4x - 3)}$

(d) $\dfrac{4x}{4x^2 - 9}$

(e) $\dfrac{4}{x^2 - 7x - 8}$

(f) $\dfrac{3x}{2x^2 - 2x - 4}$

In questions **3** to **8** express the given fraction in partial fractions.

**3** $\dfrac{3x - 1}{x^2(x - 3)}$

**4** $\dfrac{1 - 4x}{(x^2 + 1)(x + 4)}$

**5** $\dfrac{8}{(x + 3)(x - 1)^2}$

**6** $\dfrac{x^2}{(x + 1)^2(x - 1)}$

**7** $\dfrac{x}{(x - 1)(x^2 + 5)}$

**8** $\dfrac{3 - x}{(x^2 + 2)(x + 2)}$

**9** Express as the sum of a constant and partial fractions.

(a) $\dfrac{x^2}{(x + 1)(x + 2)}$    (b) $\dfrac{x^3 + 3}{x^2(x + 1)}$

**10** Express $y$ in partial fractions and hence find $\dfrac{dy}{dx}$ and $\dfrac{d^2y}{dx^2}$

(a) $y = \dfrac{2x}{(x - 1)(x - 2)}$

(b) $y = \dfrac{x}{(x + 3)(x - 2)}$

(c) $y = \dfrac{x^2 + x + 3}{(x + 1)(x + 2)(x + 3)}$

# THE BINOMIAL THEOREM FOR ANY VALUE OF $n$

We know from Pure Mathematics 1, Chapter 11 (page 107) that

$$(a + b)^n = a^n + na^{n-1}b + \frac{n(n - 1)}{2!}a^{n-2}b^2 + \frac{n(n - 1)(n - 2)}{3!}a^{n-3}b^3 + \ldots + b^n$$

for any positive integer value of $n$. This is a finite series because it ends with $b^n$.

If we write down the first few terms of the expansion of $(a + b)^n$ when $n = \frac{1}{2}$

we get

$$(a + b)^{\frac{1}{2}} = a^{\frac{1}{2}} + \frac{1}{2}a^{\left(\frac{1}{2}-1\right)}b + \frac{\frac{1}{2}\left(\frac{1}{2} - 1\right)}{2!}a^{\left(\frac{1}{2}-2\right)}b^2 + \frac{\frac{1}{2}\left(\frac{1}{2} - 1\right)\left(\frac{1}{2} - 2\right)}{3!}a^{\left(\frac{1}{2}-3\right)}b^3 + \ldots$$

It is clear that this series is not finite because the powers of $a$ can never reach zero, so the powers of $b$ continue to increase. Also, there will never be a factor equal to zero in the coefficients because there are no factors in the numerators that are positive integers.

For the series to have a sum to infinity equal to $(a + b)^{\frac{1}{2}}$, the terms need to approach zero.

Writing $(a + b)^{\frac{1}{2}}$ as $a^{\frac{1}{2}}\left(1 + \dfrac{b}{a}\right)^{\frac{1}{2}}$, the expansion becomes

$$a^{\frac{1}{2}}\left(1 + \frac{b}{a}\right)^{\frac{1}{2}} = a^{\frac{1}{2}}\left(1 + \frac{1}{2}\left(\frac{b}{a}\right) + \frac{\left(\frac{1}{2} - 1\right)}{2!}\left(\frac{b}{a}\right)^2 + \frac{\left(\frac{1}{2} - 1\right)\left(\frac{1}{2} - 2\right)}{3!}\left(\frac{b}{a}\right)^3 + \ldots\right)$$

For the terms in this series to approach zero, $\left|\dfrac{b}{a}\right| < 1$

Dividing both sides by $a^{\frac{1}{2}}$ and replacing $\dfrac{b}{a}$ by $x$ gives, for any real value of $n$

$$(1 + x)^n = 1 + nx + \frac{n(n - 1)}{2!}x^2 + \frac{n(n - 1)(n - 2)}{3!}x^3 \ldots$$

provided that $|x| < 1$

This is the general form of the binomial theorem for values of $n$ that are not positive integers, and it is important to note:

- the series does not end, it carries on to infinity
- the series is valid only for values $-1 < x < 1$

The worked examples show how to expand powers of expressions other than $(1 + x)$.

## Examples 11e

1 Expand each of the following functions as a series of ascending powers of $x$ up to and including the term in $x^3$, stating the set of values of $x$ for which each expansion is valid.

(a)  $(1 + x)^{\frac{1}{2}}$

(b)  $(1 - 2x)^{-3}$

(c)  $(2 - x)^{-2}$

For $|x| < 1$

$$(1 + x)^n = 1 + nx + \frac{n(n - 1)}{2!}x^2 + \frac{n(n - 1)(n - 2)}{3!}x^3 + \dots \qquad [1]$$

(a)  Replacing $n$ by $\frac{1}{2}$ in [1] gives

$$(1 + x)^{\frac{1}{2}} = 1 + \frac{1}{2}x + \frac{\frac{1}{2}\left(\frac{1}{2} - 1\right)}{2!}x^2 + \frac{\frac{1}{2}\left(\frac{1}{2} - 1\right)\left(\frac{1}{2} - 2\right)}{3!}x^3 + \dots$$

$$= 1 + \frac{1}{2}x + \frac{\frac{1}{2}\left(-\frac{1}{2}\right)}{2!}x^2 + \frac{\frac{1}{2}\left(-\frac{1}{2}\right)\left(-\frac{3}{2}\right)}{3!}x^3 + \dots$$

$$= 1 + \frac{x}{2} - \frac{x^2}{8} + \frac{x^3}{16} - \dots \text{ for } |x| < 1$$

(b)  Replacing $n$ by $-3$ and $x$ by $-2x$ in [1] gives

$$(1 - 2x)^{-3} = 1 + (-3)(-2x) + \frac{(-3)(-4)}{2!}(-2x)^2 + \frac{(-3)(-4)(-5)}{3!}(-2x)^3 + \dots$$

$$= 1 + 6x + 24x^2 + 80x^3 + \dots$$

provided that $|2x| < 1$, i.e. $-\frac{1}{2} < x < \frac{1}{2}$

(c)  $(2 - x)^{-2} = 2^{-2}(1 - \frac{1}{2}x)^{-2}$

Replacing $n$ by $-2$ and $x$ by $-\frac{1}{2}x$ in [1] gives

$$(2 - x)^{-2} = \frac{1}{4}\left[1 + (-2)\left(-\frac{1}{2}x\right) + \frac{(-2)(-3)}{2!}\left(-\frac{1}{2}x\right)^2 + \frac{(-2)(-3)(-4)}{3!}\left(-\frac{1}{2}x\right)^3 + \dots\right]$$

$$= \frac{1}{4}\left(1 + x + \frac{3}{4}x^2 + \frac{1}{2}x^3 + \dots\right)$$

$$= \frac{1}{4} + \frac{1}{4}x + \frac{3}{16}x^2 + \frac{1}{8}x^3 + \dots$$

The expansion of $\left(1 - \frac{1}{2}x\right)^{-2}$ is valid for $\left|\frac{1}{2}x\right| < 1$, i.e. for $-2 < x < 2$

Therefore the expansion $(2 - x)^{-\frac{1}{2}}$ also is valid for $-2 < x < 2$

2 Express $\dfrac{5}{(1 + 3x)(1 - 2x)}$ in partial fractions.

Hence expand $\dfrac{5}{(1 + 3x)(1 - 2x)}$ as a series of ascending powers of $x$, giving the first four

terms and the range of values of $x$ for which the expansion is valid.

Expressing $\dfrac{5}{(1 + 3x)(1 - 2x)}$ in partial fractions gives

$$\frac{5}{(1 + 3x)(1 - 2x)} = \frac{3}{(1 + 3x)} + \frac{2}{(1 - 2x)} = 3(1 + 3x)^{-1} + 2(1 - 2x)^{-1}$$

Now $(1 + x)^{-1} = 1 - x + x^2 - x^3 + \ldots$ for $-1 < x < 1$

Replacing $x$ by $3x$ gives

$$(1 + 3x)^{-1} = 1 - 3x + (3x)^2 - (3x)^3 + \ldots$$
$$= 1 - 3x + 9x^2 - 27x^3 + \ldots \text{ for } -1 < 3x < 1$$

Also    $(1 - x)^{-1} = 1 + x + x^2 + \ldots$ and replacing $x$ by $2x$ gives

$$(1 - 2x)^{-1} = 1 + (2x) + (2x)^2 + (2x)^3 + \ldots$$
$$= 1 + 2x + 4x^2 + 8x^3 + \ldots \text{ for } -1 < -2x < 1$$

Hence    $\dfrac{5}{(1 + 3x)(1 - 2x)} = 3(1 + 3x)^{-1} + 2(1 - 2x)^{-1}$

$$= (3 + 2) + (-9 + 4)x + (27 + 8)x^2 + (-81 + 16)x^3 + \ldots$$

provided that $-\frac{1}{3} < x < \frac{1}{3}$    *and*    $-\frac{1}{2} < x < \frac{1}{2}$

Therefore the first four terms of the series are $5 - 5x + 35x^2 - 65x^3$. The expansion is valid for the range of values of $x$ satisfying both $-\frac{1}{3} < x < \frac{1}{3}$ and $-\frac{1}{2} < x < \frac{1}{2}$

i.e. for $-\frac{1}{3} < x < \frac{1}{3}$

**3** Expand $\sqrt{\dfrac{1 + x}{1 - 2x}}$ as a series of ascending powers of $x$ up to and including the term containing $x^2$

$$\sqrt{\frac{1 + x}{1 - 2x}} \equiv (1 + x)^{\frac{1}{2}}(1 - 2x)^{-\frac{1}{2}}$$

$$(1 + x)^{\frac{1}{2}} = \left[1 + \tfrac{1}{2}x + \frac{\left(\frac{1}{2}\right)\left(-\frac{1}{2}\right)}{2!}x^2 + \ldots\right] \text{ for } -1 < x < 1$$

and $(1 - 2x)^{-\frac{1}{2}} = \left[1 + \left(-\tfrac{1}{2}\right)(-2x) + \frac{\left(-\frac{1}{2}\right)\left(-\frac{3}{2}\right)}{2!}(-2x)^2 + \ldots\right]$ for $-1 < 2x < 1$

Hence    $\sqrt{\dfrac{1 + x}{1 - 2x}} \equiv (1 + x)^{\frac{1}{2}}(1 - 2x)^{-\frac{1}{2}}$

$$= \left(1 + \tfrac{1}{2}x - \tfrac{1}{8}x^2 + \ldots\right)\left(1 + x + \tfrac{3}{2}x^2 + \ldots\right)$$

$$= 1 + \left(\tfrac{1}{2}x + x\right) + \left(\tfrac{1}{2}x^2 - \tfrac{1}{8}x^2 + \tfrac{3}{2}x^2\right) + \ldots$$

$$= 1 + \tfrac{3}{2}x + \tfrac{15}{8}x^2 + \ldots$$

We ignore all terms in this product containing $x^3$ or higher powers of $x$.

provided that $-1 < x < 1$ *and* $-\frac{1}{2} < x < \frac{1}{2}$, i.e. $-\frac{1}{2} < x < \frac{1}{2}$

## Exercise 11e

Expand the following functions as series of ascending powers of $x$ up to and including the term in $x^3$. In each case give the range of values of $x$ for which the expansion is valid.

**1** $(1 - 2x)^{\frac{1}{2}}$

**2** $(1 + 5x)^{-2}$

**3** $\left(1 - \frac{1}{2}x\right)^{-3}$

**4** $(1 + x)^{\frac{3}{2}}$

**5** $(3 + x)^{-1}$

**6** $\left(1 + \frac{x}{2}\right)^{-\frac{1}{2}}$

**7** $\dfrac{1}{(1 - x)^2}$

**8** $\sqrt{\dfrac{1}{1 + x}}$

**9** $(1 + x)\sqrt{1 - x}$

**10** $\dfrac{x + 2}{x - 1}$

**11** $\dfrac{2 - x}{\sqrt{1 - 3x}}$

**12** $\dfrac{1}{(2 - x)(1 + 2x)}$

**13** $\sqrt{\dfrac{1 + x}{1 - x}}$

**14** $\left(1 + \dfrac{x^2}{9}\right)^{-1}$

**15** $\dfrac{x}{(1 + x)(1 - 2x)}$

**16** $\left(1 + \dfrac{1}{x}\right)^{-1}$ $\left[\text{Hint: } \left(1 + \frac{1}{x}\right)^{-1} \equiv \left(\frac{x + 1}{x}\right)^{-1} \equiv \frac{x}{1 + x}\right]$

**17** Expand $\left(1 + \dfrac{1}{p}\right)^{-3}$ as a series of descending powers of $p$, as far as and including the term containing $p^{-4}$. State the range of values of $p$ for which the expansion is valid.

$\left(\text{Hint: Replace } x \text{ by } \frac{1}{p} \text{ in } (1 + x)^{-3}\right)$

**18** Expand $\sqrt{\dfrac{1 + 2x}{1 - 2x}}$ as a series of ascending powers of $x$ up to and including the term in $x^2$.

**19** If $x$ is so small that $x^2$ and higher powers of $x$ may be neglected, show that

$$\frac{1}{(x - 1)(x + 2)} \approx -\frac{1}{2} - \frac{1}{4}x$$

**20** By neglecting $x^3$ and higher powers of $x$, find a quadratic function that approximates to the function $\dfrac{1 - 2x}{\sqrt{1 + 2x}}$ for small values of $x$.

**21** Find a quadratic function that approximates to

$$f(x) = \frac{1}{\sqrt[3]{(1 - 3x)^2}}$$

for values of $x$ that are small enough for $x^3$ and higher powers to be neglected.

**22** Use partial fractions and the binomial series to show that

$$\frac{3}{(1 - 2x)(2 - x)} \approx \frac{3}{2} + \frac{15}{4}x$$

**23** If terms containing $x^4$ and higher powers of $x$ can be neglected, show that

$$\frac{2}{(x + 1)(x^2 + 1)} \approx 2(1 - x)$$

**24** Show that

$$\frac{12}{(3 + x)(1 - x)^2} \approx 4 + \frac{20}{3}x + \frac{88}{9}x^2$$

provided that $x$ is small enough to neglect powers higher than 2.

**25** If $x$ is very small, show that

$$\frac{1}{(3 - x)^3} \approx \frac{1}{729}(27 + 27x + 18x^2 + 10x^3)$$

# 12 Integration 2

## After studying this chapter you should be able to

- integrate rational functions by means of decomposition into partial fractions
- recognise an integrand of the form $\dfrac{f'(x)}{f(x)}$, and integrate, for example, $\dfrac{x}{x^2 + 1}$ or $\tan x$
- recognise when an integrand can usefully be regarded as a product, and use integration by parts to integrate, for example, $x \sin 2x$, $x^2 e^x$ or $\ln x$
- use a given substitution to simplify and evaluate either a definite or an indefinite integral.

## INTEGRATING RATIONAL FUNCTIONS

An integral such as $\displaystyle\int \dfrac{x}{(x - 1)(x + 2)} \, dx$ can be found by decomposing the integrand into partial fractions,

i.e. $\dfrac{x}{(x - 1)(x + 2)} \equiv \dfrac{A}{x - 1} + \dfrac{B}{x + 2} \quad \Rightarrow \quad x \equiv A(x + 2) + B(x - 1)$

$x = 1$ gives $A = \frac{1}{3}$ and $x = -2$ gives $B = \frac{2}{3}$

Therefore $\displaystyle\int \dfrac{x}{(x - 1)(x + 2)} \, dx = \int \dfrac{1}{3(x - 1)} \, dx + \int \dfrac{2}{3(x + 2)} \, dx$

$$= \tfrac{1}{3} \ln |x - 1| + \tfrac{2}{3} \ln |x + 2| + k$$

### Exercise 12a

Express each function in partial fractions and hence find the integral of each function with respect to $x$.

1  $\dfrac{2}{x(x + 1)}$

2  $\dfrac{4}{(x - 2)(x + 2)}$

3  $\dfrac{x}{(x - 1)(x + 1)}$

4  $\dfrac{x - 1}{x(x + 2)}$

5  $\dfrac{x - 1}{(x - 2)(x - 3)}$

6  $\dfrac{1}{x(x - 1)(x + 1)}$

7  $\dfrac{x}{x^2 - 1}$

8  $\dfrac{2}{x^2 - 1}$

9  $\dfrac{2x}{x^2 - 5x + 6}$

10  $\dfrac{2x - 3}{x^2 - 5x + 6}$

Use partial fractions to evaluate

11  $\displaystyle\int_3^4 \dfrac{x}{(x - 2)(x + 2)} \, dx$

12  $\displaystyle\int_0^1 \dfrac{2x}{(x + 1)(x - 3)} \, dx$

13  $\displaystyle\int_{-1}^1 \dfrac{5}{x^2 + x - 6} \, dx$

14  $\displaystyle\int_1^2 \dfrac{x + 1}{x(x + 4)} \, dx$

## INTEGRATING FUNCTIONS OF THE FORM $\dfrac{f'(x)}{f(x)}$

We know that $\dfrac{d}{dx} \ln f(x) = \dfrac{f'(x)}{f(x)}$

Therefore an integral of the form $\displaystyle\int \dfrac{f'(x)}{f(x)} \, dx$ can be recognised as $\ln |f(x)| + k$

For example, $\displaystyle\int \dfrac{x}{x^2 + 2} \, dx = \dfrac{1}{2} \int \dfrac{2x}{x^2 + 2} \, dx = \dfrac{1}{2} \ln |x^2 + 2| + k$

## Examples 12b

**1** Find $\displaystyle\int \frac{x^2}{1 + x^3}\, dx$

$$\int \frac{x^2}{1 + x^3}\, dx = \frac{1}{3}\int \frac{3x^2}{1 + x^3}\, dx$$

This integral is of the form $\displaystyle\int \frac{f'(x)}{f(x)}\, dx$ so we use recognition.

$$= \frac{1}{3}\ln |1 + x^3| + K$$

**2** By writing $\tan x$ as $\dfrac{\sin x}{\cos x}$ find $\displaystyle\int \tan x\, dx$

$$\int \tan x\, dx = \int \frac{\sin x}{\cos x}\, dx = -\int \frac{f'(x)}{f(x)}\, dx \text{ where } f(x) = \cos x$$

so $\displaystyle\int \frac{\sin x}{\cos x}\, dx = -\ln |\cos x| + K$

$\therefore \displaystyle\int \tan x\, dx = K - \ln |\cos x|$

## Exercise 12b

Integrate each function with respect to $x$

**1** $\dfrac{\cos x}{4 + \sin x}$

**2** $\dfrac{e^x}{3e^x - 1}$

**3** $\dfrac{x}{(1 - x^2)}$

**4** $\dfrac{\cos x}{\sin x}$

**5** $\dfrac{x^3}{1 + x^4}$

**6** $\dfrac{2x + 3}{x^2 + 3x - 4}$

**7** $\dfrac{1}{x \ln x}$  i.e.  $\dfrac{\frac{1}{x}}{\ln x}$

**8** $\dfrac{2x}{1 - x^2}$

**9** $\dfrac{\sin x}{\cos x}$

**10** $\dfrac{\sec x \tan x}{4 + \sec x}$

**11** $\dfrac{x - 1}{x(x - 2)}$

**12** $\dfrac{e^x - 1}{(e^x - x)}$

Find the value of

**13** $\displaystyle\int_1^2 \frac{2x + 1}{x^2 + x}\, dx$

**14** $\displaystyle\int_0^1 \frac{x}{x^2 + 1}\, dx$

## INTEGRATION BY PARTS (INTEGRATING PRODUCTS)

We use a formula to differentiate a product, $uv$, where $u$ and $v$ are functions of $x$,

i.e  $\dfrac{d}{dx}(uv) = v\dfrac{du}{dx} + u\dfrac{dv}{dx}$

Integrating with respect to $x$ gives

$$\int \frac{d}{dx}(uv)\, dx = \int v\frac{du}{dx}\, dx + \int u\frac{dv}{dx}\, dx \quad \Rightarrow \quad uv = \int v\frac{du}{dx}\, dx + \int u\frac{dv}{dx}\, dx$$

Rearranging gives

$$\int v \frac{du}{dx}\, dx = uv - \int u \frac{dv}{dx}\, dx$$

This version of the formula can be used to integrate a product where $v$ and $\frac{du}{dx}$ are the two factors of the product.

On the right-hand side of the formula, we see that one factor, $\frac{du}{dx}$, has to be integrated and the other factor, $v$, has to be differentiated. When both factors can be integrated, choose $v$ to be the factor whose differential is the simpler. When only one factor can be integrated, the other factor is chosen as $v$. The formula cannot be used when neither factor can be integrated.

---

## Examples 12c

1  Integrate $xe^x$ with respect to $x$

Taking $\qquad v = x$ and $\dfrac{du}{dx} = e^x$

gives $\qquad \dfrac{dv}{dx} = 1$ and $u = e^x$

> Both $x$ and $e^x$ can be integrated, but the differential of $x$ is simpler than the differential of $e^x$.

Then $\qquad \int v \dfrac{du}{dx}\, dx = uv - \int u \dfrac{dv}{dx}\, dx$

gives $\qquad \int xe^x\, dx = (e^x)(x) - \int (e^x)(1)\, dx$

$\qquad\qquad\qquad = xe^x - e^x + K$

2  Find $\int x^4 \ln x\, dx$

Taking $\qquad v = \ln x$ and $\dfrac{du}{dx} = x^4$

gives $\qquad \dfrac{dv}{dx} = \dfrac{1}{x}$ and $u = \dfrac{1}{5}x^5$

> Because $\ln x$ can be differentiated but *not integrated*, we must use $v = \ln x$

Integrating by parts then gives

$$\int x^4 \ln x\, dx = \left(\tfrac{1}{5}x^5\right)(\ln x) - \int\left(\tfrac{1}{5}x^5\right)\left(\tfrac{1}{x}\right) dx = \tfrac{1}{5}x^5 \ln x - \tfrac{1}{5}\int x^4\, dx$$

$$\Rightarrow \quad \int x^4 \ln x\, dx = \tfrac{1}{5}x^5 \ln x - \tfrac{1}{25}x^5 + K$$

3  Find $\int \ln x\, dx$

Taking $\ln x$ as the product of $\ln x$ and $1$

gives $\qquad\qquad v = \ln x$ and $\dfrac{du}{dx} = 1$

$\Rightarrow \qquad\qquad \dfrac{dv}{dx} = \dfrac{1}{x}$ and $u = x$

Then $\qquad \int v \dfrac{du}{dx}\, dx = uv - \int u \dfrac{dv}{dx}\, dx$

becomes    $\int \ln x \, dx = x \ln x - \int x \left( \frac{1}{x} \right) dx = x \ln x - x + K$

i.e.    $\int \ln x \, dx = x(\ln x - 1) + K$

**4**  Find $\int x^2 \sin x \, dx$

Taking    $v = x^2$ and $\frac{du}{dx} = \sin x$

gives    $\frac{dv}{dx} = 2x$ and $u = -\cos x$

Then    $\int v \frac{du}{dx} \, dx = uv - \int u \frac{dv}{dx} \, dx$

gives    $\int x^2 \sin x \, dx = (-\cos x)(x^2) - \int (-\cos x)(2x) \, dx$

$= -x^2 \cos x + 2 \int x \cos x \, dx$    [1]

We need to use integration
by parts again to find
$\int x \cos x \, dx$

Taking    $v = x$ and $\frac{du}{dx} = \cos x$

gives    $\frac{dv}{dx} = 1$ and $u = \sin x$

Then    $\int x \cos x \, dx = (\sin x)(x) - \int (\sin x)(1) \, dx = x \sin x + \cos x + K$

Hence equation [1] becomes

$\int x^2 \sin x \, dx = -x^2 \cos x + 2x \sin x + 2 \cos x + K$

## Exercise 12c

Integrate the following functions with respect to $x$

1  $x \cos x$

2  $x e^{2x}$

3  $x^3 \ln 3x$

4  $x e^{-x}$

5  $3x \sin x$

6  $(1 - x)e^x$

7  $e^x \sin x$

8  $x e^{(x-1)}$

9  $(1 - 2x)e^{2x}$

10  $\ln 2x$

11  $e^x(x + 1)$

12  $x(1 + x)^7$

13  $x \sin \left( x + \frac{1}{6}\pi \right)$

14  $x \cos nx$

15  $x^n \ln x$

16  $3x \cos 2x$

17  $(3x - 2)\cos x$

18  $\frac{1}{x} \ln x$

19  $2x\sqrt{e^x}$

20  $x^{\frac{1}{2}} \ln x$

21  $x^2 e^x$

22  $x^2 \sin x$

23  $x^2 e^{4x}$

24  $e^{2x} \cos x$

## Definite integration by parts

The term $uv$ in the formula

$$\int v \frac{du}{dx} \, dx = uv - \int u \frac{dv}{dx} \, dx$$

is fully integrated. Therefore in a definite integration, $uv$ must be *evaluated between the appropriate boundaries*.

i.e.  $\int_a^b v \frac{du}{dx} \, dx = \left[ uv \right]_a^b - \int_a^b u \frac{dv}{dx} \, dx$

---

### Example 12d

Evaluate $\int_0^1 xe^x \, dx$

$$\int xe^x \, dx = \int v \frac{du}{dx} \, dx = uv - \int u \frac{dv}{dx} \, dx$$

using  $v = x$ and $\dfrac{du}{dx} = e^x$

gives  $\int_0^1 xe^x \, dx = \left[ xe^x \right]_0^1 - \int_0^1 e^x \, dx$

$$= \left[ xe^x \right]_0^1 - \left[ e^x \right]_0^1$$

$$= (e^1 - 0) - (e^1 - e^0)$$

$$= e - e + 1$$

i.e.  $\int_0^1 xe^x \, dx = 1$

---

### Exercise 12d

Evaluate

1  $\int_0^{\frac{\pi}{2}} x \sin x \, dx$

2  $\int_1^2 x^5 \ln x \, dx$

3  $\int_0^1 (x + 1)e^x \, dx$

4  $\int_0^{\pi} 2x \cos x \, dx$

5  $\int_1^2 x\sqrt{x - 1} \, dx$

6  $\int_1^2 x \ln x \, dx$

7  $\int_0^1 \ln (1 + x) \, dx$

8  $\int_0^1 (1 - x)e^{-x} \, dx$

9  $\int_0^{\frac{\pi}{4}} x \sin 2x \, dx$

## Differentiation and integration of inverse trigonometric functions

This section is not in the syllabus for P3.

We can use implicit differentiation to find $\dfrac{d}{dx} (\sin^{-1} x)$

Starting with $y = \sin^{-1} x$ gives $\sin y = x$

Differentiating with respect to $x$ gives $\cos y \dfrac{dy}{dx} = 1 \quad \Rightarrow \quad \dfrac{dy}{dx} = \dfrac{1}{\cos y}$

Using the identity $\cos^2 y + \sin^2 y = 1$ gives $\cos y = \sqrt{1 - \sin^2 y} = \sqrt{1 - x^2}$

(We use the positive root only because the gradient of $f(x) = \sin^{-1} x$ is positive for all values of $x$ in its domain.)

$\therefore \quad \dfrac{dy}{dx} = \dfrac{1}{\sqrt{1 - x^2}} \qquad$ i.e. $\qquad \dfrac{d}{dx}(\sin^{-1} x) = \dfrac{1}{\sqrt{1 - x^2}}$

and it follows that $\qquad \displaystyle\int \dfrac{1}{\sqrt{1 - x^2}}\, dx = \sin^{-1} x + K$

We can now use integration by parts to find $\displaystyle\int \sin^{-1} x\, dx$.

---

**Example 12e**

Find $\displaystyle\int \sin^{-1} x\, dx$

Taking $v = \sin^{-1} x$ and $\dfrac{du}{dx} = 1$

$$\int \sin^{-1} x\, dx = x \sin^{-1} x - \int \dfrac{x}{\sqrt{1 - x^2}}\, dx$$

$$= x \sin^{-1} x + \sqrt{1 - x^2} + K$$

---

**Exercise 12e**

1  Differentiate each function with respect to $x$.

  (a) $\cos^{-1} x$

  (b) $\sin^{-1} 2x$

  (c) $\tan^{-1} x$

2  Integrate each function with respect to $x$.

  (a) $\cos^{-1} x$

  (b) $\sin^{-1} 2x$

  (c) $\tan^{-1} x$

## INTEGRATION USING SUBSTITUTION

For a function $g(u)$ where $u$ is a function of $x$, using the chain rule gives

$$\dfrac{d}{dx} g(u) = \dfrac{du}{dx} g'(u) \text{ or } g'(u) \dfrac{du}{dx}$$

Therefore $\qquad \displaystyle\int g'(u) \dfrac{du}{dx}\, dx = g(u) + K \qquad$ [1]

We also know that $\qquad \displaystyle\int g'(u)\, du = g(u) + K \qquad$ [2]

Comparing [1] and [2] gives

$$\int g'(u) \frac{du}{dx} \, dx = \int g'(u) \, du$$

Replacing $g'(u)$ by $f(u)$ gives

$$\int f(u) \frac{du}{dx} \, dx = \int f(u) \, du$$

i.e. $$\ldots \frac{du}{dx} \, dx \equiv \ldots du \qquad\qquad\qquad [3]$$

Therefore integrating (a function of $u$) $\frac{du}{dx}$ with respect to $x$, is *equivalent* to integrating (the same function of $u$) with respect to $u$.

This means that the relationship in [3] is not an equation nor is it an identity. It is a pair of equivalent operations.

For example, to find $\int 2x(x^2 + 1)^5 \, dx$,

using the substitution $u = x^2 + 1$ gives

$$\int (x^2 + 1)^5 2x \, dx = \int u^5 (2x) \, dx$$

Then $\frac{du}{dx} = 2x$ and as $\ldots \frac{du}{dx} \, dx \equiv \ldots du$

we have $\ldots 2x \, dx \equiv \ldots du$

i.e. $\int (x^2 + 1)^5 2x \, dx = \int u^5 \, du$

$$= \tfrac{1}{6} u^6 + K = \tfrac{1}{6}(x^2 + 1)^6 + K$$

---

### Examples 12f

1  Use the substitution $u = x^3 + 5$ to find $\int x^2 \sqrt{x^3 + 5} \, dx$

$$u = x^3 + 5 \quad \Rightarrow \quad \frac{du}{dx} = 3x^2$$

$$\Rightarrow \qquad\qquad \ldots du \equiv \ldots 3x^2 \, dx$$

$$\therefore \qquad \int x^2 \sqrt{x^3 + 5} \, dx = \tfrac{1}{3} \int (x^3 + 5)^{\frac{1}{2}} (3x^2 \, dx) = \tfrac{1}{3} \int u^{\frac{1}{2}} \, du$$

$$= \left(\tfrac{1}{3}\right)\left(\tfrac{2}{3}\right) u^{\frac{3}{2}} + K = \tfrac{2}{9}(x^3 + 5)^{\frac{3}{2}} + K$$

2  Use the substitution $u = \sin x$ to find $\int \cos x \sin^3 x \, dx$

$$u = \sin x \quad \Rightarrow \quad \ldots du \equiv \ldots \cos x \, dx$$

$$\therefore \quad \int \cos x \sin^3 x \, dx = \int (\sin x)^3 \cos x \, dx = \int u^3 \, du$$

$$= \tfrac{1}{4} u^4 + K = \tfrac{1}{4} \sin^4 x + K$$

3 Use the substitution $u = 1 - e^x$ to find $\int \dfrac{e^x}{(1 - e^x)^2}\, dx$

$$u = 1 - e^x \quad \Rightarrow \quad \dots du \equiv \dots -e^x\, dx$$

So $\quad \displaystyle\int \dfrac{e^x}{(1 - e^x)^2}\, dx = \int \dfrac{-1}{u^2}\, du = \dfrac{1}{u} + K$

$$\int \dfrac{e^x}{(1 - e^x)^2}\, dx = \dfrac{1}{1 - e^x} + K$$

4 Use the substitution $u^2 = 1 + x$ to find $\int \dfrac{x}{\sqrt{1 + x}}\, dx$

$$u^2 = 1 + x \quad \Rightarrow \quad \dots 2u\, du = \dots dx \quad \text{and} \quad x = u^2 - 1$$

So $\quad \displaystyle\int \dfrac{x}{\sqrt{1 + x}}\, dx = \int \dfrac{u^2 - 1}{u} \times 2u\, du = \int 2(u^2 - 1)\, du = \dfrac{2}{3}u^3 - 2u + K$

$\therefore \quad \displaystyle\int \dfrac{x}{\sqrt{1 + x}}\, dx = \dfrac{2}{3}(1 + x)^{\frac{3}{2}} - 2(1 + x)^{\frac{1}{2}} + K$

5 Use the substitution $u = \sin x$ to find $\int \sec x\, dx$

$$u = \sin x \quad \Rightarrow \quad \dots du = \dots \cos x\, dx \quad \Rightarrow \quad \dots dx = \dots \dfrac{1}{\cos x}\, du$$

$\therefore \quad \displaystyle\int \sec x\, dx = \int \dfrac{1}{\cos^2 x}\, du = \int \dfrac{1}{1 - u^2}\, du \quad (\text{using } \sin^2 x + \cos^2 x \equiv 1)$

Using partial fractions

$$\int \dfrac{1}{1 - u^2}\, du = \int \dfrac{1}{2}\left(\dfrac{1}{u - 1} - \dfrac{1}{u + 1}\right) du = \dfrac{1}{2}\ln|u - 1| - \dfrac{1}{2}\ln|u + 1| + K$$

$$= \dfrac{1}{2}\ln\left|\dfrac{\sin x - 1}{\sin x + 1}\right| + K$$

## Exercise 12f

Find the following integrals by using the substitution given.

1 $\displaystyle\int x(x^2 - 3)^4\, dx;$  $\quad u = x^2 - 3$

2 $\displaystyle\int x\sqrt{1 - x^2}\, dx;$  $\quad u = 1 - x^2$

3 $\displaystyle\int \cos 2x\,(\sin 2x + 3)^2\, dx;$  $\quad u = \sin 2x + 3$

4 $\displaystyle\int x^2\,(1 - x^3)\, dx;$  $\quad u = 1 - x^3$

5 $\displaystyle\int e^x\sqrt{1 + e^x}\, dx;$  $\quad u = 1 + e^x$

6 $\displaystyle\int \cos x \sin^4 x\, dx;$  $\quad u = \sin x$

7 $\displaystyle\int \sec^2 x \tan^3 x\, dx;$  $\quad u = \tan x$

8 $\displaystyle\int x^n(1 + x^{n+1})^2\, dx;$  $\quad u = 1 + x^{n+1}$

9 $\displaystyle\int \operatorname{cosec}^2 x \cot^2 x\, dx;$  $\quad u = \cot x$

10 $\displaystyle\int \sqrt{x}\,\sqrt{1 + x^{\frac{3}{2}}}\, dx;$  $\quad u = 1 + x^{\frac{3}{2}}$

11 $\displaystyle\int (x + 1)(x + 3)^5\, dx;$  $\quad x + 3 = u$

12 $\displaystyle\int \frac{x}{\sqrt{3-x}}\,dx;$ $\qquad$ $3 - x = u^2$ $\qquad$ 19 $\displaystyle\int x^3(x^4 + 4)^2\,dx;$ $\qquad$ $u = x^4 + 4$

13 $\displaystyle\int x\sqrt{x+1}\,dx;$ $\qquad$ $x + 1 = u^2$ $\qquad$ 20 $\displaystyle\int e^x\,(1 - e^x)^3\,dx;$ $\qquad$ $u = 1 - e^x$

14 $\displaystyle\int \frac{2x+1}{(x-3)^6}\,dx;$ $\qquad$ $x - 3 = u$ $\qquad$ 21 $\displaystyle\int \sin\theta\sqrt{1 - \cos\theta}\,d\theta;$ $\qquad$ $u^2 = 1 - \cos\theta$

15 $\displaystyle\int 2x\sqrt{3x-4}\,dx;$ $\qquad$ $3x - 4 = u^2$ $\qquad$ 22 $\displaystyle\int (x+1)\sqrt{x^2 + 2x + 3}\,dx;$ $\quad$ $u^2 = x^2 + 2x + 3$

16 $\displaystyle\int 2x(1-x)^7\,dx;$ $\qquad$ $u = 1 - x$ $\qquad$ 23 $\displaystyle\int xe^{x^2 + 1}\,dx;$ $\qquad$ $u = x^2 + 1$

17 $\displaystyle\int \frac{x+3}{(4-x)^5}\,dx$ $\qquad$ $u = 4 - x$ $\qquad$ 24 $\displaystyle\int \sec^2 x(1 + \tan x)\,dx;$ $\qquad$ $u = 1 + \tan x$

18 $\displaystyle\int \frac{\sin x}{\sqrt{7 + \cos x}}\,dx;$ $\qquad$ $u^2 = 7 + \cos x$ $\qquad$ 25 $\displaystyle\int \operatorname{cosec} x\,dx;$ $\qquad$ $u = \cos x$

## DEFINITE INTEGRATION WITH A CHANGE OF VARIABLE

When using a substitution, $u = f(x)$, to find a definite integral, use $u = f(x)$ to change the limits of the integral from values of $x$ to the equivalent values of $u$.

---

### Example 12g

By using the substitution $u = x^3 + 1$, evaluate $\displaystyle\int_0^1 x^2\sqrt{x^3 + 1}\,dx$

If $\quad u = x^3 + 1 \quad$ then $\quad \ldots du \equiv \ldots 3x^2\,dx$

and $\quad \begin{cases} x = 0 & \Rightarrow \quad u = 1 \\ x = 1 & \Rightarrow \quad u = 2 \end{cases}$

Hence $\quad \displaystyle\int_0^1 x^2\sqrt{x^3 + 1}\,dx = \frac{1}{3}\int_1^2 \sqrt{u}\,du$

$$= \frac{1}{3}\left[\frac{2}{3}u^{\frac{3}{2}}\right]_1^2 = \frac{2}{9}\left(2\sqrt{2} - 1\right)$$

---

### Exercise 12g

Evaluate

1 $\displaystyle\int_0^1 x\,e^{x^2}\,dx;$ $\qquad$ $u = x^2$ $\qquad$ 6 $\displaystyle\int_1^2 x(1 + 2x^2)\,dx;$ $\qquad$ $u = 1 + 2x^2$

2 $\displaystyle\int_0^{\frac{\pi}{2}} \cos x \sin^4 x\,dx;$ $\qquad$ $u = \sin^4 x$ $\qquad$ 7 $\displaystyle\int_2^3 (x-1)\,e^{(x^2 - 2x)}\,dx;$ $\qquad$ $u = x^2 - 2x$

3 $\displaystyle\int_1^2 \frac{1}{x}\ln x\,dx;$ $\qquad$ $u = \ln x$ $\qquad$ 8 $\displaystyle\int_0^{\frac{\pi}{6}} \cos x\,(1 + \sin^2 x)\,dx;$ $\quad$ $u = \sin x$

4 $\displaystyle\int_1^2 x^2(x^3 - 1)^4\,dx;$ $\qquad$ $u = x^3 - 1$ $\qquad$ 9 $\displaystyle\int_1^3 \frac{1}{x}(\ln x)^2\,dx;$ $\qquad$ $u = \ln x$

5 $\displaystyle\int_0^{\frac{\pi}{4}} (\sec^2 x)e^{\tan x}\,dx;$ $\qquad$ $u = \tan x$ $\qquad$ 10 $\displaystyle\int_0^{\sqrt{3}} x\sqrt{1 + x^2}\,dx;$ $\qquad$ $u^2 = 1 + x^2$

## Summary

To integrate a fraction,

when it is of the form $\dfrac{f'(x)}{f(x)}$, $\displaystyle\int \dfrac{f'(x)}{f(x)}\,dx = \ln|f(x)|$

when the denominator has linear and/or quadratic factors, decompose to give partial fractions which will usually be of the form $\dfrac{f'(x)}{f(x)}$

use a substitution when it is given.

To integrate a product,

when one factor can be integrated and the other factor can be differentiated, use

integration by parts, i.e. $\displaystyle\int v\,\dfrac{du}{dx}\,dx = uv - \int u\,\dfrac{dv}{dx}\,dx$

use a substitution when it is given.

## Mixed exercise 12

Integrate the following functions with respect to $x$.

**1** $x(1 + x^2)^4$;     $u = 1 + x^2$

**2** $xe^{-3x}$

**3** $x^2 e^{2x}$

**4** $\dfrac{x + 3}{x + 2}$

**5** $\dfrac{x^2}{(x^3 + 1)^2}$;     $u = x^3 + 1$

**6** $\dfrac{3}{(x - 4)(x - 1)}$

**7** $\dfrac{(x + 1)}{x(2x + 1)}$

**8** $\dfrac{x - 1}{(x^2 + 1)}$;     $u = x^2 + 1$

**9** $\dfrac{\sin x}{\sqrt{\cos x}}$;     $u^2 = \cos x$

Evaluate

**10** $\displaystyle\int_{\frac{\pi}{2}}^{\pi} \left(\sin \tfrac{1}{2}x + \cos 2x\right) dx$

**11** $\displaystyle\int_{2}^{5} x\sqrt{x - 1}\,dx$;     $u = x - 1$

**12** $\displaystyle\int_{0}^{\frac{\pi}{4}} \tan x\,dx$

**13** $\displaystyle\int_{1}^{2} x\sqrt{5 - x^2}\,dx$;     $u^2 = 5 - x^2$

**14** $\displaystyle\int_{4}^{6} \dfrac{5}{x^2 - x - 6}\,dx$

**15** $\displaystyle\int_{1}^{2} \dfrac{2x}{x^2 + 1}\,dx$

**16** $\displaystyle\int_{-2}^{-1} \dfrac{2 - x}{x(1 - x)}\,dx$

# 13 Differential equations

## After studying this chapter you should be able to

- formulate a simple statement involving a rate of change as a differential equation, including the introduction if necessary of a constant of proportionality
- find by integration a general form of solution for a first order differential equation in which the variables are separable
- use an initial condition to find a particular solution
- interpret the solution of a differential equation in the context of a problem being modelled by the equation.

## DIFFERENTIAL EQUATIONS

An equation in which at least one term contains $\frac{dy}{dx}$, $\frac{d^2y}{dx^2}$ etc. is called a *differential equation*.

If it contains only $\frac{dy}{dx}$ it is a first order differential equation, for example, $x + 2\frac{dy}{dx} = 3y$ is a first order differential equation.

This example is a *linear* differential equation because $\frac{dy}{dx}$ is not raised to a power higher than 1.

A differential equation is a relationship between two variables. The same relationship can often be expressed in a form that does not contain a differential coefficient.

For example $\frac{dy}{dx} = 2x$ and $y = x^2 + K$ express the same relationship between $x$ and $y$, but $\frac{dy}{dx} = 2x$ is a differential equation whereas $y = x^2 + K$ is not.

Converting a differential equation into a direct one is called *solving the differential equation*.

## FIRST ORDER DIFFERENTIAL EQUATIONS WITH SEPARABLE VARIABLES

Integrating both sides of the equation $3y\frac{dy}{dx} = 5x^2$ gives

$$\int 3y\frac{dy}{dx}\,dx = \int 5x^2\,dx \qquad\qquad [1]$$

We know that $\ldots \frac{dy}{dx}\,dx \equiv \ldots dy$

so $\qquad \int 3y\,dy = \int 5x^2\,dx$ [2]

Temporarily removing the integral signs from this equation gives

$\qquad 3y\,dy = 5x^2\,dx \qquad [3]$

This can be found from equation [1] by *separating the variables*; separate $dy$ from $dx$ then collect all the terms involving $y$ with $dy$ and on the other side all the terms involving $x$ with $dx$.

The 'equation' [3] has no meaning and should not be written down. It only gives a quick mental way of changing the differential equation [1] to [2] where each side can be integrated separately.

Integrating each side of equation [2] gives

$$\frac{3y^2}{2} = \frac{5}{3}x^3 + A$$

(Adding a constant of integration to one side is enough.)

### Example 13a

Solve the differential equation $\dfrac{1}{x}\dfrac{dy}{dx} = \dfrac{2y}{x^2 + 1}$

$$\frac{1}{x}\frac{dy}{dx} = \frac{2y}{x^2 + 1} \qquad \Rightarrow \qquad \frac{1}{y}\frac{dy}{dx} = \frac{2x}{x^2 + 1}$$

So, after separating the variables we have

$$\int \frac{1}{y}\,dy = \int \frac{2x}{x^2 + 1}\,dx$$

$$\frac{1}{\text{\ding{172}}x}\frac{dy}{dx} = \frac{2\text{\ding{172}}y}{x^2 + 1}$$

$$\Rightarrow \qquad \ln|y| = \ln|x^2 + 1| + A$$

### Exercise 13a

Solve each differential equation.

**1**   $y\dfrac{dy}{dx} = \sin x$

**2**   $x^2\dfrac{dy}{dx} = y^2$

**3**   $\dfrac{1}{x}\dfrac{dy}{dx} = \dfrac{1}{y^2 - 2}$

**4**   $\tan y\dfrac{dy}{dx} = \dfrac{1}{x}$

**5**   $\dfrac{dy}{dx} = y^2$

**6**   $\dfrac{1}{x}\dfrac{dy}{dx} = \dfrac{1}{1 - x^2}$

**7**   $(x - 3)\dfrac{dy}{dx} = y$

**8**   $\tan y\dfrac{dx}{dy} = 4$

**9**   $u\dfrac{du}{dv} = v + 2$

**10**   $e^x\dfrac{dy}{dx} = \dfrac{x}{y}$

**11**   $\dfrac{dv}{du} = \dfrac{v + 1}{u + 2}$

**12**   $y(x + 1) = (x^2 + 2x)\dfrac{dy}{dx}$

**13**   $v^2\dfrac{dv}{dt} = (2 + t)^3$

**14**   $x\dfrac{dy}{dx} = \dfrac{1}{y} + y$

**15**   $y\sin^3 x\dfrac{dy}{dx} = \cos x$

**16**   $\dfrac{uv}{u - 1} = \dfrac{du}{dv}$

**17**   $e^x\dfrac{dy}{dx} = e^{y - 1}$

### Calculation of the constant of integration

We need extra information to find the constant of integration.

For example, the point $(0, 2)$ is on a curve and the equation of a curve satisfies the differential equation $2\dfrac{dy}{dx} = \dfrac{\cos x}{y}$

To find the equation of the curve, we first need to solve the differential equation.

Separating the variables gives

$$\int 2y\,dy = \int \cos x\,dx \qquad \Rightarrow \qquad y^2 = A + \sin x$$

When      $x = 0, y = 2$ so $4 = A$

Therefore     $y^2 = 4 + \sin x$

## Example 13b

The gradient of a curve is proportional to the product of the $x$- and $y$-coordinates. The curve passes through the points $(2, 1)$ and $(4, e^2)$. Find its equation.

$$\frac{dy}{dx} \propto xy \qquad \Rightarrow \qquad \frac{dy}{dx} = kxy \text{ where } k \text{ is a constant of proportion.}$$

$$\therefore \quad \int \frac{1}{y} \, dy = \int kx \, dx \qquad \Rightarrow \qquad \ln |y| = \tfrac{1}{2} kx^2 + A$$

There are *two* unknown constants this time so we need two extra pieces of information.

$$y = 1 \text{ when } x = 2 \qquad \Rightarrow \qquad \ln 1 = 2k + A$$
$$\ln 1 = 0 \text{ so } A + 2k = 0$$

and $\quad y = e^2$ when $x = 4 \qquad \Rightarrow \qquad \ln e^2 = 8k + A$
$$\ln e^2 = 2 \text{ so } A + 8k = 2$$

Solving these equations for $A$ and $k$ we get $k = \tfrac{1}{3}$ and $A = -\tfrac{2}{3}$

$\therefore \quad$ the equation of the curve is $\ln |y| = \tfrac{1}{6} x^2 - \tfrac{2}{3}$

## Exercise 13b

Solve each of the following differential equations.

1  $y^2 \dfrac{dy}{dx} = x^2 + 1$ and $y = 1$ when $x = 2$

2  $e^t \dfrac{ds}{dt} = \sqrt{s}$ and $s = 4$ when $t = 0$

3  $\dfrac{y}{x} \dfrac{dy}{dx} = \dfrac{y^2 - 1}{x^2 - 1}$ and $y = 3$ when $x = 2$

4  A curve passes through the origin and its gradient function is $2x - 1$. Find the equation of the curve and sketch it.

5  A curve for which $e^{-x} \dfrac{dy}{dx} = 1$ passes through the point $(0, -1)$. Find the equation of the curve.

6  A curve passes through the points $(1, 2)$ and $\left(\tfrac{1}{5}, -10\right)$ and its gradient is inversely proportional to $x^2$. Find the equation of the curve.

7  Given that $y = 2$ when $x = 1$, find the coordinates of the point where the curve represented by $\dfrac{2y}{3} \dfrac{dy}{dx} = e^{-3x}$ crosses the $y$-axis.

8  Find the equation of the curve whose gradient function is $\dfrac{y + 1}{x^2 - 1}$ and that passes through the point $(-3, 1)$.

9  The gradient function of a curve is proportional to $x + 3$. The curve passes through the origin and the point $(2, 8)$. Find its equation.

10  Solve the differential equation
$(1 + x^2) \dfrac{dy}{dx} - y(y + 1)x = 0$, given that $y = 1$ when $x = 0$

11  Solve the differential equation $\dfrac{dy}{dx} = 3x^2y^2$ given that $y = 1$ when $x = 0$

12  $\dfrac{dy}{dx} = x(y^2 + 1)$ and $y = 0$ when $x = 2$. Solve the differential equation.

## GENERAL RATES OF INCREASE

We know that

$$\frac{dy}{dx} \text{ represents the rate at which } y \text{ increases compared with } x$$

When the variation in one quantity, $p$, depends on the changing value of another quantity, $q$, then the rate of increase of $p$ compared with $q$ can be expressed as $\frac{dp}{dq}$.

There are many situations where such relationships exist, e.g.

1) Liquid expands when it is heated. When $V$ is the volume of a quantity of liquid and $T$ is the temperature, then the rate at which the volume increases with temperature can be written $\frac{dV}{dT}$

2) The profit, $P$, made by a company selling cars depends upon the number, $n$, of cars sold. So $\frac{dP}{dn}$ represents the rate of increase of profit compared with sales.

## NATURAL OCCURRENCE OF DIFFERENTIAL EQUATIONS

Differential equations often come from a physical situation when it is modelled mathematically.

For example:

1) A body falls from rest in a medium that causes the velocity to decrease at a rate proportional to the velocity.

   The velocity is *decreasing* with time, so its rate of increase is *negative*.

   Using $v$ for velocity and $t$ for time, the rate of change of velocity can be written as $-\frac{dv}{dt}$, so

   $$-\frac{dv}{dt} \propto v$$

   Then the motion of the body satisfies the differential equation

   $$-\frac{dv}{dt} = kv$$

2) When yeast cells grow in a culture, the number of cells increases in proportion to the number already formed.

   So $n$, the number of cells at a particular time $t$, is such that $\frac{dn}{dt} \propto n$ and can be found from the differential equation

   $$\frac{dn}{dt} = kn$$

### Exercise 13c

In questions **1** to **3** form, but *do not solve*, the differential equations representing the given data.

1  A body moves with a velocity $v$ which is inversely proportional to its displacement $s$ from a fixed point.

2  The rate at which the height $h$ of a plant increases is proportional to the natural logarithm of the difference between its present height and its final height $H$.

3 In a community the number, $n$, of people suffering from a disease is $N_1$ at a particular time. The disease then becomes epidemic so that the number of sick people increases at a rate proportional to $n$, until the total number of sufferers is $N_2$. The rate of increase then becomes inversely proportional to $n$ until $N_3$ people have the disease. After this, the total number of sick people decreases at a constant rate. Write down the differential equation governing the incidence of the disease

(a)  for $N_1 \leqslant n \leqslant N_2$

(b)  for $N_2 \leqslant n \leqslant N_3$

(c)  for $n \geqslant N_3$

## Solving naturally occurring differential equations

Naturally occurring differential equations often involve a constant of proportion. The initial solution will involve this constant and a constant of integration. Therefore two pieces of information are needed to find the values of these constants.

### Examples 13d

1  A particle moves in a straight line with an acceleration that is inversely proportional to its velocity. (Acceleration is the rate of increase of velocity.)

(a)  Form a differential equation to represent this data.

(b)  Given that the acceleration is $2\,\mathrm{m\,s^{-2}}$ when the velocity is $5\,\mathrm{m\,s^{-1}}$, solve the differential equation.

(a)  Using $\dfrac{\mathrm{d}v}{\mathrm{d}t}$ for acceleration gives $\dfrac{\mathrm{d}v}{\mathrm{d}t} \propto \dfrac{1}{v} \quad \Rightarrow \quad \dfrac{\mathrm{d}v}{\mathrm{d}t} = \dfrac{k}{v}$

(b)  $v = 5$ when $\dfrac{\mathrm{d}v}{\mathrm{d}t} = 2$, so $2 = \dfrac{k}{5} \quad \Rightarrow \quad k = 10$

$\therefore \quad \dfrac{\mathrm{d}v}{\mathrm{d}t} = \dfrac{10}{v}$

Separating the variables gives $\displaystyle\int v\,\mathrm{d}v = \int 10\,\mathrm{d}t \quad \Rightarrow \quad \tfrac{1}{2}v^2 = 10t + A$

> The value of $A$ cannot be found because we do not have enough information.

2  The rate at which the atoms in a mass of radioactive material are disintegrating is proportional to $N$, the number of atoms present at any time. Initially the number of atoms is $M$.

(a)  Form and solve the differential equation that represents this information.

(b)  Given that half of the original mass disintegrates in 152 days, evaluate the constant of proportion in the differential equation.

(a)  The rate at which the atoms are disintegrating is $-\dfrac{\mathrm{d}N}{\mathrm{d}t}$

$\therefore \quad -\dfrac{\mathrm{d}N}{\mathrm{d}t} = kN$

Separating the variable gives $\displaystyle\int \dfrac{1}{N}\,\mathrm{d}N = -\int k\,\mathrm{d}t$

$\therefore \quad \ln N = -kt + \ln B \quad \Rightarrow \quad \ln \dfrac{N}{B} = -kt \quad \Rightarrow \quad N = Be^{-kt}$

When $t = 0$, $N = M \quad \Rightarrow \quad B = M$

$\therefore \quad N = Me^{-kt}$

> Writing the constant of integration as ln B instead of A makes it easier to simplify the equation.

(b)  When $N = \tfrac{1}{2}M, t = 152$

$\therefore \quad \tfrac{1}{2}M = Me^{-152k} \quad \Rightarrow \quad \ln\left(\tfrac{1}{2}\right) = -152k$

$\therefore \quad 152k = \ln 2 \quad \Rightarrow \quad k = 0.00456 \text{ (3 s.f.)}$

## Exercise 13d

1 Grain is pouring from a container on to a barn floor where it forms a conical pile whose height $h$ is increasing at a rate that is inversely proportional to $h^3$. The initial height of the pile is $h_0$ and the height doubles after a time $T$. Find, in terms of $T$, the time after which its height has grown to $3h_0$

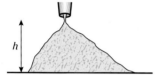

2 The gradient of any point of a curve is proportional to the square root of the $x$-coordinate. Given that the curve passes through the point $(1, 2)$ and at that point the gradient is 0.6, form and solve the differential equation representing the given relationship. Show that the curve passes through the point $(4, 4.8)$ and find the gradient at this point.

3 The number of bacteria in a liquid is growing at a rate proportional to the number of bacteria present at any time. Initially there are 100 bacteria.

(a) Form a differential equation that models the growth in the number of bacteria.

(b) The number of bacteria increases by 50% in 10 hours. Find the number of hours for the bacteria to double from the initial number of 100.

4 In a chemical reaction, a substance is transformed into a compound. The mass of the substance after time $t$ is $m$. The substance is being transformed at a rate that is proportional to the mass of the substance at that time. Given that the original mass is 50 g and that 20 g is transformed after 200 seconds

(a) form and solve the differential equation relating $m$ and $t$

(b) find the mass of the substance transformed in 300 seconds.

5 The rate of decrease of the temperature of a liquid is proportional to the amount by which this temperature is greater than the temperature of its surroundings. (This is known as Newton's Law of Cooling.) Take $\theta$ as the difference in temperature at any time $t$, and $80°$ as the initial difference.

(a) Show that $\theta = 80e^{-kt}$

A pan of water at $65\,°C$ is standing in a kitchen whose temperature is $15\,°C$.

(b) Show that, after cooling for $t$ minutes, the water temperature, $\phi$, can be modelled by the equation

$\phi = 15 + 50e^{-kt}$ where $k$ is a constant.

(c) Given that after 10 minutes the temperature of the water has fallen to $50\,°C$, find the value of $k$.

(d) Find the temperature after 15 minutes.

6 A virus has infected the population of rabbits on an island. The growth in the number of rabbits infected is proportional to the number already infected. Initially 20 rabbits were infected.

(a) Form a differential equation that models the growth in the number infected.

(b) Thirty days after the initial number of infections, 60 rabbits were infected. After how many further days does the model predict that 200 rabbits will be infected?

# 14 Vectors

## After studying this chapter you should be able to

- understand the significance of all the symbols used when the equation of a straight line is expressed in the form $r = a + tb$
- determine whether two lines are parallel, intersect or are skew
- find the angle between two lines, and the point of intersection of two lines when it exists
- understand the significance of all the symbols used when the equation of a plane is expressed in either of the forms $ax + by + cz = d$ or $(r - a).n = 0$
- use equations of lines and planes to solve problems concerning distances, angles and intersections, and in particular:
    find the equation of a line or a plane, given sufficient information
    determine whether a line lies in a plane, is parallel to a plane, or intersects a plane, and find the point of intersection of a line and a plane when it exists
    find the line of intersection of two non-parallel planes
    find the perpendicular distance from a point to a plane, and from a point to a line
    find the angle between two planes, and the angle between a line and a plane.

## THE EQUATION OF A STRAIGHT LINE

A line is exactly located in space if

(a) it has a known direction and passes through a known fixed point, or

(b) it passes through two known fixed points.

## A line with known direction passing through a fixed point

Consider a line that is parallel to a vector **b** and that passes through a fixed point $A$ with position vector **a**.

If **r** is the position vector, $\overrightarrow{OP}$, of a point $P$ then

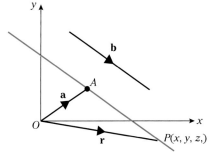

$$P \text{ is a point on this line} \quad \Leftrightarrow \quad \overrightarrow{AP} = t\,\mathbf{b}$$

where $t$ is a variable scalar, i.e. a parameter.

Now $\qquad \overrightarrow{OP} = \overrightarrow{OA} + \overrightarrow{AP}$

i.e. $\qquad \mathbf{r} = \mathbf{a} + t\mathbf{b}$

i.e. $\qquad P \text{ is on the line} \quad \Leftrightarrow \quad \mathbf{r} = \mathbf{a} + t\mathbf{b}$

For each value of the parameter $t$ this equation gives the position vector of one point on the line and it is called the vector equation of the line.

**a** is the position vector of any point on the line so **a** can have many different values. This means that the equation of the line is not unique.

For example, the line whose vector equation is

$$\mathbf{r} = (5\mathbf{i} - 2\mathbf{j} + 4\mathbf{k}) + t(2\mathbf{i} - \mathbf{j} + 3\mathbf{k})$$

is *parallel* to the vector $2\mathbf{i} - \mathbf{j} + 3\mathbf{k}$ and passes through the point whose position vector is $5\mathbf{i} - 2\mathbf{j} + 4\mathbf{k}$.

Sometimes the equation of a line is given as, for example, $\mathbf{r} = (t + 1)\mathbf{i} - t\mathbf{j} + (2 - 3t)\mathbf{k}$

This equation can be rearranged as $\mathbf{r} = \mathbf{i} + 2\mathbf{k} + t(\mathbf{i} - \mathbf{j} - 3\mathbf{k})$. So this line passes through the point with position vector $\mathbf{i} + 2\mathbf{k}$ and is parallel to $\mathbf{i} - \mathbf{j} - 3\mathbf{k}$.

## A line passing through two given points

To find the equation of a line $l$ passing through two given points with position vectors $\mathbf{a}$ and $\mathbf{c}$, we can use one of the points as the fixed point, and, as the diagram shows, the difference between the vectors is the vector parallel to the line. So an equation of the line $l$ is $\mathbf{r} = \mathbf{a} + t(\mathbf{a} - \mathbf{c})$

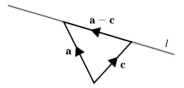

For example, to find an equation of the line passing through the points $(1, 2, 1)$ and $(3, -2, -1)$ we can use the position vector of $(1, 2, 1)$ as $\mathbf{a}$, and the difference in the position vectors of $(1, 2, 1)$ and $(3, -2, -1)$ as a vector parallel to the line.

Therefore an equation of this line is $\mathbf{r} = \mathbf{i} + 2\mathbf{j} + \mathbf{k} + t\{(\mathbf{i} + 2\mathbf{j} + \mathbf{k}) - (3\mathbf{i} - 2\mathbf{j} - \mathbf{k})\}$

$\Rightarrow \quad \mathbf{r} = \mathbf{i} + 2\mathbf{j} + \mathbf{k} + t\{(-2\mathbf{i} + 4\mathbf{j} + 2\mathbf{k})$

---

### Examples 14a

1  Find a vector equation of the line that passes through the point with position vector $2\mathbf{i} - \mathbf{j} + 4\mathbf{k}$ and is parallel to the vector $\mathbf{i} + \mathbf{j} - 2\mathbf{k}$

> The vector equation of a line is $\mathbf{r} = \mathbf{a} + t\mathbf{b}$ where $\mathbf{a}$ is the position vector of a point on the line and $\mathbf{b}$ is parallel to the line. For this line, $\mathbf{a} = 2\mathbf{i} - \mathbf{j} + 4\mathbf{k}$ and $\mathbf{b} = \mathbf{i} + \mathbf{j} - 2\mathbf{k}$

A vector equation of the line is $\mathbf{r} = 2\mathbf{i} - \mathbf{j} + 4\mathbf{k} + t(\mathbf{i} + \mathbf{j} - 2\mathbf{k})$

2  Find a vector equation for the line through the points $A(3, 4, -7)$ and $B(1, -1, 6)$

> To find a vector equation of a line we need a point on the line (we can use either $A$ or $B$) and a vector parallel to the line.

As $A$ and $B$ are on the line, $\overrightarrow{AB}$ is parallel to the line and $\overrightarrow{AB} = \overrightarrow{OB} - \overrightarrow{OA}$

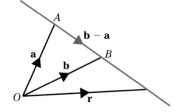

$\overrightarrow{OA} = \mathbf{a} = 3\mathbf{i} + 4\mathbf{j} - 7\mathbf{k}$ and $\overrightarrow{OB} = \mathbf{b} = \mathbf{i} - \mathbf{j} + 6\mathbf{k}$

$\therefore \quad \overrightarrow{AB} = \mathbf{b} - \mathbf{a} = (\mathbf{i} - \mathbf{j} + 6\mathbf{k}) - (3\mathbf{i} + 4\mathbf{j} - 7\mathbf{k})$

$= -2\mathbf{i} - 5\mathbf{j} + 13\mathbf{k}$

A vector equation of the line is

$\mathbf{r} = 3\mathbf{i} + 4\mathbf{j} - 7\mathbf{k} + t(-2\mathbf{i} - 5\mathbf{j} + 13\mathbf{k})$

> This equation is not unique; we could have used $\mathbf{b}$ instead of $\mathbf{a}$ and also any multiple of $\mathbf{b} - \mathbf{a}$, all of which are parallel to the line, e.g. $\mathbf{r} = \mathbf{i} - \mathbf{j} + \mathbf{k} + t(2\mathbf{i} + 5\mathbf{j} - 13\mathbf{k})$ is an equally valid vector equation for this line.

3  State whether or not the lines with equations $\mathbf{r} = 2\mathbf{i} - 3\mathbf{j} + 2\mathbf{k} + \lambda(\mathbf{i} - \mathbf{j} + 4\mathbf{k})$ and $\mathbf{r} = (3 - \mu)\mathbf{i} - (3 - \mu)\mathbf{j} + (2 - 4\mu)\mathbf{k}$ are parallel.

> To determine whether the lines are parallel we need to find a vector in the direction of each line. This can be 'read' from the equation of the first line but the equation of the second line needs rearranging first.

$\mathbf{r} = 2\mathbf{i} - 3\mathbf{j} + 2\mathbf{k} + \lambda(\mathbf{i} - \mathbf{j} + 4\mathbf{k})$ is in the direction of the vector $(\mathbf{i} - \mathbf{j} + 4\mathbf{k})$.

$\mathbf{r} = (3 - \mu)\mathbf{i} - (3 - \mu)\mathbf{j} + (2 - 4\mu)\mathbf{k}$

$= 3\mathbf{i} - 3\mathbf{j} + 2\mathbf{k} + \mu(-\mathbf{i} + \mathbf{j} - 4\mathbf{k})$ so is in the direction of the vector $(-\mathbf{i} + \mathbf{j} - 4\mathbf{k})$.

$-\mathbf{i} + \mathbf{j} - 4\mathbf{k} = -(\mathbf{i} - \mathbf{j} + 4\mathbf{k})$  $\therefore$  the lines are parallel.

4  Find the coordinates of the point where the line $\mathbf{r} = 2\mathbf{i} - 3\mathbf{j} + 2\mathbf{k} + s(\mathbf{i} - \mathbf{j} + 4\mathbf{k})$ cuts the *xy*-plane.

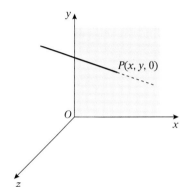

The *z*-coordinate of any point *P* on the *xy*-plane is zero,
i.e. *OP* has zero **k** component.
We can identify the **k** component if we rearrange the equation.

Rearranging the equation of the line as
$\mathbf{r} = (2 + s)\mathbf{i} + (-3 - s)\mathbf{j} + (2 + 4s)\mathbf{k}$ shows that it cuts
the *xy*-plane where $2 + 4s = 0$, i.e. where $s = -\frac{1}{2}$

When  $s = -\frac{1}{2}$, $\mathbf{r} = \frac{3}{2}\mathbf{i} - \frac{7}{2}\mathbf{j}$

$\therefore$  the line cuts the *xy*-plane at the point $\left(\frac{3}{2}, -\frac{7}{2}, 0\right)$.

5  The position vectors of the points *A* and *B* are $(3\mathbf{i} + 2\mathbf{j} - 2\mathbf{k})$ and $(2\mathbf{i} + \mathbf{j} + 5\mathbf{k})$ respectively.

The point *C* is the midpoint of *AB* and *D* is the point with position vector $(\mathbf{i} - 4\mathbf{j} + \mathbf{k})$.

Find the equation of the line through *C* and *D*.

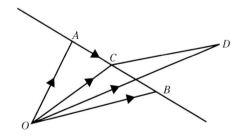

The equation of the line through *C* and *D* is $\mathbf{r} = \overrightarrow{OC} + t\overrightarrow{CD}$

$\overrightarrow{OC} = \overrightarrow{OA} + \frac{1}{2}\overrightarrow{AB}$ and $\overrightarrow{AB} = \overrightarrow{OB} - \overrightarrow{OA}$

$\overrightarrow{AB} = (2\mathbf{i} + \mathbf{j} + 5\mathbf{k}) - (3\mathbf{i} + 2\mathbf{j} - 2\mathbf{k}) = -\mathbf{i} - \mathbf{j} + 7\mathbf{k}$

$\therefore$  $\overrightarrow{OC} = (3\mathbf{i} + 2\mathbf{j} - 2\mathbf{k}) + \frac{1}{2}(-\mathbf{i} - \mathbf{j} + 7\mathbf{k}) = \frac{1}{2}(5\mathbf{i} + 3\mathbf{j} + 3\mathbf{k})$

$\overrightarrow{CD} = \overrightarrow{OD} - \overrightarrow{OC} = (\mathbf{i} - 4\mathbf{j} + \mathbf{k}) - \frac{1}{2}(5\mathbf{i} + 3\mathbf{j} + 3\mathbf{k}) = \frac{1}{2}(-3\mathbf{i} - 11\mathbf{j} - \mathbf{k})$

$\therefore$  the equation of the line through *C* and *D* is

$\mathbf{r} = \frac{1}{2}(5\mathbf{i} + 3\mathbf{j} + 3\mathbf{k}) + t(3\mathbf{i} + 11\mathbf{j} + \mathbf{k})$

## The perpendicular distance of a point from a line

$C$ is the point with position vector $\mathbf{c}$ and $l$ is the line with equation $\mathbf{r} = \mathbf{a} + t\mathbf{b}$

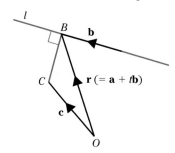

To find the perpendicular distance of $C$ from $l$, we first need to find a point $B$ on the line such that $CB$ is perpendicular to $l$.

$\mathbf{r}$ is the position vector of any point on the line. So for any point $B$ on the line

$$\overrightarrow{CB} = (\mathbf{r} - \mathbf{c}) = (\mathbf{a} + t\mathbf{b} - \mathbf{c})$$

For $\overrightarrow{CB}$ to be perpendicular to $l$, $\overrightarrow{CB}.\mathbf{b} = 0$ using the scalar product.

Therefore $(\mathbf{a} + t\mathbf{b} - \mathbf{c}).\mathbf{b} = 0$ can be solved to find the value of $t$.

Using this value of $t$, the magnitude of $\mathbf{a} + t\mathbf{b} - \mathbf{c}$ is the perpendicular distance of $C$ from $l$.

The next worked example illustrates this.

---

### Examples 14a cont.

6 Find the perpendicular distance of the point $A(1, 3, 2)$ from the line $\mathbf{r} = \begin{pmatrix} 2 \\ 5 \\ 1 \end{pmatrix} + t\begin{pmatrix} -1 \\ 2 \\ 1 \end{pmatrix}$

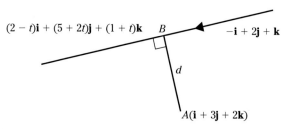

We need a point $B$ on the line so that $AB$ is perpendicular to the line; i.e. we need a value of $t$ such that $(-\mathbf{i} + 2\mathbf{j} + \mathbf{k}).\overrightarrow{AB} = 0$

$\Rightarrow \quad (-\mathbf{i} + 2\mathbf{j} + \mathbf{k}).\{(1 - t)\mathbf{i} + (2 + 2t)\mathbf{j} + (-1 + t)\mathbf{k}\} = 0$

$\Rightarrow \quad -(1 - t) + (4 + 4t) + (-1 + t) = 0 \Rightarrow t = -\frac{1}{3}$

$\therefore \quad$ $B$ is the point with position vector $\frac{7}{3}\mathbf{i} + \frac{13}{3}\mathbf{j} + \frac{2}{3}\mathbf{k}$

$\therefore \quad \overrightarrow{AB} = \frac{4}{3}\mathbf{i} + \frac{4}{3}\mathbf{j} - \frac{4}{3}\mathbf{k}$

$\qquad d = \frac{1}{3}\sqrt{16 + 16 + 16} = \frac{4\sqrt{3}}{3}$

### Exercise 14a

1  Write down a vector that is parallel to each of these lines.

   (a)  $\mathbf{r} = \mathbf{i} - 2\mathbf{j} + 4\mathbf{k} + t(2\mathbf{i} - \mathbf{j} - 5\mathbf{k})$

   (b)  $\mathbf{r} = 2\mathbf{i} - \mathbf{k} + s(3\mathbf{j} - 5\mathbf{k})$

   (c)  $\mathbf{r} = (1 - 2s)\mathbf{i} + (4s - 3)\mathbf{j} + (1 + s)\mathbf{k}$

   (d)  $\mathbf{r} = t\mathbf{i} + 3\mathbf{j} - (1 - t)\mathbf{k}$

2  Write down equations in vector form for the line through a point $A$ with position vector $\mathbf{a}$ and in the direction of vector $\mathbf{b}$ where

   (a)  $\mathbf{a} = \mathbf{i} - 3\mathbf{j} + 2\mathbf{k}$    $\mathbf{b} = 5\mathbf{i} + 4\mathbf{j} - \mathbf{k}$

   (b)  $\mathbf{a} = 2\mathbf{i} + \mathbf{j}$    $\mathbf{b} = 3\mathbf{j} - \mathbf{k}$

   (c)  $A$ is the origin    $\mathbf{b} = \mathbf{i} - \mathbf{j} - \mathbf{k}$

3  State whether or not the following pairs of lines are parallel.

   (a)  $\mathbf{r} = \mathbf{i} + \mathbf{j} - \mathbf{k} + \lambda(2\mathbf{i} - 3\mathbf{j} + \mathbf{k})$
        $\mathbf{r} = 2\mathbf{i} - 4\mathbf{j} + 5\mathbf{k} + \lambda(\mathbf{i} + \mathbf{j} - \mathbf{k})$

   (b)  $\mathbf{r} = 2\mathbf{i} - \mathbf{j} + 5\mathbf{k} + s(\mathbf{i} + \mathbf{j} - \mathbf{k})$
        $\mathbf{r} = (3 + t)\mathbf{i} + (t - 1)\mathbf{j} + (5 - t)\mathbf{k}$

   (c)  $\mathbf{r} = 2\mathbf{i} - \mathbf{j} + 4\mathbf{k} + \lambda(\mathbf{i} + \mathbf{j} + 3\mathbf{k})$
        $\mathbf{r} = \mu(2\mathbf{i} + 2\mathbf{j} + 6\mathbf{k})$

   (d)  $\mathbf{r} = \lambda(3\mathbf{i} - 3\mathbf{j} + 6\mathbf{k})$
        $\mathbf{r} = 4\mathbf{j} + \lambda(-\mathbf{i} + \mathbf{j} - 2\mathbf{k})$

4  The points $A(4, 5, 10)$, $B(2, 3, 4)$ and $C(1, 2, -1)$ are three vertices of a parallelogram $ABCD$.

   (a)  Find vector equations for the sides $AB$, $BC$ and $AD$.

   (b)  Find the perpendicular distance of the point $A$ from $BC$.

5  Write down a vector equation for the line through $A$ and $B$ if

   (a)  $\overrightarrow{AB}$ is $3\mathbf{i} + \mathbf{j} - 4\mathbf{k}$ and $\overrightarrow{OB}$ is $\mathbf{i} + 7\mathbf{j} + 8\mathbf{k}$

   (b)  $A$ and $B$ have coordinates $(1, 1, 7)$ and $(3, 4, 1)$.

   Find, in each case, the coordinates of the points where the line crosses the $xy$-plane, the $yz$-plane and the $zx$-plane.

6  Find the distance of the point with position vector $\begin{pmatrix} 2 \\ -1 \\ 1 \end{pmatrix}$ from the line $\mathbf{r} = \begin{pmatrix} 1 \\ -2 \\ 2 \end{pmatrix} + t\begin{pmatrix} 1 \\ 1 \\ 4 \end{pmatrix}$

## PAIRS OF LINES

Two lines in space may be

(a)  parallel

(b)  not parallel and intersecting

(c)  not parallel and not intersecting. Such lines are called *skew*.

### Non-parallel lines

Consider two lines whose vector equations are $\mathbf{r}_1 = \mathbf{a}_1 + t\mathbf{b}_1$ and $\mathbf{r}_2 = \mathbf{a}_2 + s\mathbf{b}_2$

If these lines intersect, there must be values of $t$ and $s$ such that

$\quad \mathbf{a}_1 + t\mathbf{b}_1 = \mathbf{a}_2 + s\mathbf{b}_2$

If no such values can be found, the lines do not intersect.

The next worked example shows how to find these values.

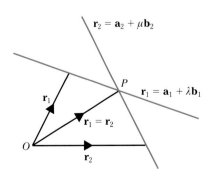

### Examples 14b

1  Find out whether the following pairs of lines are parallel, non-parallel and intersecting, or non-parallel and non-intersecting.

   (a)  $\mathbf{r}_1 = \mathbf{i} + \mathbf{j} + 2\mathbf{k} + t(3\mathbf{i} - 2\mathbf{j} + 4\mathbf{k})$ and $\mathbf{r}_2 = 2\mathbf{i} - \mathbf{j} + 3\mathbf{k} + s(-6\mathbf{i} + 4\mathbf{j} - 8\mathbf{k})$

   (b)  $\mathbf{r}_1 = \mathbf{i} - \mathbf{j} + 3\mathbf{k} + t(\mathbf{i} - \mathbf{j} + \mathbf{k})$ and $\mathbf{r}_2 = 2\mathbf{i} + 4\mathbf{j} + 6\mathbf{k} + s(2\mathbf{i} + \mathbf{j} + 3\mathbf{k})$

   (c)  $\mathbf{r}_1 = \mathbf{i} + \mathbf{k} + t(\mathbf{i} + 3\mathbf{j} + 4\mathbf{k})$ and $\mathbf{r}_2 = 2\mathbf{i} + 3\mathbf{j} + s(4\mathbf{i} - \mathbf{j} + \mathbf{k})$

   (a)  Checking first whether the lines are parallel, we compare their directions.

        The first line is parallel to $3\mathbf{i} - 2\mathbf{j} + 4\mathbf{k}$

        The second line is parallel to $-6\mathbf{i} + 4\mathbf{j} - 8\mathbf{k} = -2(3\mathbf{i} - 2\mathbf{j} + 4\mathbf{k})$

        Therefore these two lines are parallel.

   (b)  In this case the directions of the lines are $\mathbf{i} - \mathbf{j} + \mathbf{k}$ and $2\mathbf{i} + \mathbf{j} + 3\mathbf{k}$.

        These are not equal, so these two lines are not parallel.

        If the lines intersect it will be at a point where $\mathbf{r}_1 = \mathbf{r}_2$, i.e. where

$$(1 + t)\mathbf{i} - (1 + t)\mathbf{j} + (3 + t)\mathbf{k} = 2(1 + s)\mathbf{i} + (4 + s)\mathbf{j} + (6 + 3s)\mathbf{k}$$

        Equating the coefficients of $\mathbf{i}$ and $\mathbf{j}$, we have

$$1 + t = 2(1 + s) \text{ and } -(1 + t) = 4 + s$$

        Hence    $s = -2,\ t = -3$

        With these values for $t$ and $s$, the coefficients of $\mathbf{k}$ are:

        first line     $3 + t = 0$  } equal values.
        second line    $6 + 3s = 0$

        So    $\mathbf{r}_1 = \mathbf{r}_2$ when $t = -3$ and $s = -2$

        Therefore the lines *do* intersect at the point with position vector

$$(1 - 3)\mathbf{i} - (1 - 3)\mathbf{j} + (3 - 3)\mathbf{k}, \quad (t = -3 \text{ in } \mathbf{r}_1) \quad \text{i.e.} \quad -2\mathbf{i} + 2\mathbf{j}.$$

   (c)  The directions of these two lines are not equal so the lines are not parallel.

        If the lines intersect it will be where $\mathbf{r}_1 = \mathbf{r}_2$, i.e. where

$$(1 + t)\mathbf{i} + 3t\mathbf{j} + (1 + 4t)\mathbf{k} = (2 + 4s)\mathbf{i} + (3 - s)\mathbf{j} + s\mathbf{k}$$

        Equating the coefficients of $\mathbf{i}$ and $\mathbf{j}$, we have

        $1 + t = 2 + 4s$  }  $\Rightarrow$   $s = 0,$   $t = 1$
        $3t = 3 - s$

        With these values of $\lambda$ and $\mu$, the coefficients of $\mathbf{k}$ become

        first line      $1 + 4t = 5$  } unequal values.
        second line     $s = 0$

        So there are no values of $t$ and $s$ for which $\mathbf{r}_1 = \mathbf{r}_2$ and these lines do not intersect and are skew.

2 Find the acute angle between the lines
$$\mathbf{r} = 2\mathbf{i} - \mathbf{j} + 4\mathbf{k} + \lambda(\mathbf{i} + \mathbf{j} - 2\mathbf{k}) \text{ and } \mathbf{r} = (3 - 2\mu)\mathbf{i} + (4 - 5\mu)\mathbf{j} + (-7 + 13\mu)\mathbf{k}$$

If $\theta$ is the angle between the lines, then

$$(\mathbf{i} + \mathbf{j} - 2\mathbf{k}).(-2\mathbf{i} - 5\mathbf{j} + 13\mathbf{k})$$

$$= |\mathbf{i} + \mathbf{j} - 2\mathbf{k}| \, |-2\mathbf{i} - 5\mathbf{j} + 13\mathbf{k}| \cos\theta$$

$$\Rightarrow \quad -33 = (\sqrt{6})(\sqrt{198}) \cos\theta$$

$$\Rightarrow \quad \cos\theta = -\frac{33}{6\sqrt{33}}, \quad \text{i.e.} \quad \theta = 16.8° \text{ (1 d.p.)}$$

> The angle between the lines is the angle between their directions, i.e. between the vectors $\mathbf{i} + \mathbf{j} - 2\mathbf{k}$ and $-2\mathbf{i} - 5\mathbf{j} + 13\mathbf{k}$, which we can find by using the scalar product.

## Exercise 14b

1 Find whether the following pairs of lines are parallel, intersecting or skew. When the lines intersect, state the position vector of the common point, and find the angle between the lines.

(a) $\mathbf{r} = \mathbf{i} - \mathbf{j} + \mathbf{k} + \lambda(3\mathbf{i} - 4\mathbf{j} + \mathbf{k})$,
$\mathbf{r} = \mu(-9\mathbf{i} + 12\mathbf{j} - 3\mathbf{k})$

(b) $\mathbf{r} = (4 - t)\mathbf{i} + (8 - 2t)\mathbf{j} + (3 - t)\mathbf{k}$,
$\mathbf{r} = (7 + 6s)\mathbf{i} + (6 + 4s)\mathbf{j} + (5 + 5s)\mathbf{k}$

(c) $\mathbf{r} = \mathbf{i} + 3\mathbf{k} + \lambda(2\mathbf{i} + \mathbf{j} + \mathbf{k})$,
$\mathbf{r} = 2\mathbf{i} - \mathbf{j} + \mathbf{k} + \mu(\mathbf{i} - 2\mathbf{j})$

2 Two lines which intersect have equations
$$\mathbf{r} = 2\mathbf{i} + 9\mathbf{j} + 13\mathbf{k} + \lambda(\mathbf{i} + 2\mathbf{j} + 3\mathbf{k})$$
and $\quad \mathbf{r} = a\mathbf{i} + 7\mathbf{j} - 2\mathbf{k} + \mu(-\mathbf{i} + 2\mathbf{j} - 3\mathbf{k})$
Find the value of $a$, the position vector of the point of intersection, and the angle between the lines.

3 Show that the lines
$$\mathbf{r} = 2\mathbf{i} - \mathbf{j} + \mathbf{k} + \lambda(\mathbf{i} - 2\mathbf{j} + 2\mathbf{k})$$
and $\quad \mathbf{r} = \mathbf{i} - 3\mathbf{j} + 4\mathbf{k} + \mu(2\mathbf{i} + 3\mathbf{j} - 6\mathbf{k})$
are skew.

## EQUATION OF A PLANE

A plane can be defined in several ways, for example:
(a) one and only one plane can be drawn through three non-collinear points, therefore three given points specify a particular plane
(b) one and only one plane can be drawn to contain two concurrent lines, therefore two given concurrent lines specify a particular plane
(c) one and only one plane can be drawn perpendicular to a given direction at a given distance from the origin, therefore the normal to a plane and the distance of the plane from the origin specify a particular plane
(d) one and only one plane can be drawn through a given point and perpendicular to a given direction, therefore a point on the plane and a normal to the plane specify a particular plane.

We will use the way a plane is specified in (d) to derive the vector equation of a plane.

### The vector equation of a plane

The point $A$, whose position vector is $\mathbf{a}$, is in the plane.
The vector $\mathbf{n}$ is perpendicular to the plane.

The vector $\mathbf{r}$ is the position vector of any point in the plane.

The vector $\mathbf{r} - \mathbf{a}$ is in the plane and so it is perpendicular to $\mathbf{n}$.

Therefore the scalar product of $\mathbf{r} - \mathbf{a}$ and $\mathbf{n}$ is zero,

i.e. $\quad (\mathbf{r} - \mathbf{a}).\mathbf{n} = 0$

This is called the vector equation of the plane.

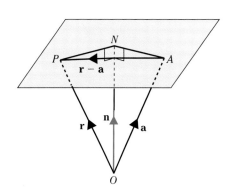

## The Cartesian equation of a plane

**r** is the position vector of any point on the plane that is perpendicular to **n**.

The scalar product of **r** and the unit vector $\hat{\mathbf{n}}$ is $r \cos P\hat{O}N$. This is the length of $ON$, i.e. it is the distance, $d$, of the plane from the origin.

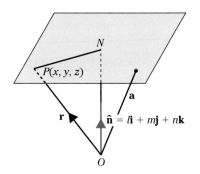

When $\mathbf{r} = x\mathbf{i} + y\mathbf{j} + z\mathbf{k}$ and $\hat{\mathbf{n}} = l\mathbf{i} + m\mathbf{j} + n\mathbf{k}$,  $\mathbf{r}.\hat{\mathbf{n}} = d$

so $(x\mathbf{i} + y\mathbf{j} + z\mathbf{k}).(l\mathbf{i} + m\mathbf{j} + n\mathbf{k}) = d$

$\Rightarrow$  $lx + my + nz = d$

i.e.  **$lx + my + nz = d$ is the Cartesian equation of the plane**

> **where $d$ is the distance of the plane from the origin and $l\mathbf{i} + m\mathbf{j} + n\mathbf{k}$ is the unit vector perpendicular to the plane.**

This equation can be multiplied by any constant.

Therefore an equation such as $3x - 2y + z = 7$ is the equation of a plane.

and $3\mathbf{i} - 2\mathbf{j} + \mathbf{z}$ is a vector perpendicular to the plane.

When we divide by $\sqrt{3^2 + (-2)^2 + 1^2} = \sqrt{14}$ we change $3\mathbf{i} - 2\mathbf{j} + \mathbf{z}$ to a unit vector,

therefore dividing both sides by $\sqrt{14}$ gives $\dfrac{3}{\sqrt{14}}x - \dfrac{2}{\sqrt{14}}y + \dfrac{1}{\sqrt{14}}z = \dfrac{7}{\sqrt{14}}$

And we can now see that this plane is a distance of $\dfrac{7}{\sqrt{14}}$ units from the origin.

---

### Examples 14c

1  Find the Cartesian equation of the plane containing the points $A(0, 1, 1)$, $B(2, 1, 0)$ and $C(-2, 0, 3)$.

The equation of the plane is $ax + by + cz = d$.
So the coordinates of $A$, $B$ and $C$ satisfy this equation

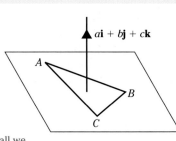

$\Rightarrow$  $b + c = d$  [1]  using A

$\Rightarrow$  $2a + b = d$  [2]  using B

$\Rightarrow$  $-2a + 3c = d$  [3]  using C

These three equations are enough to find $a$, $b$ and $c$ in terms of $d$. This is all we need because any multiple of $ax + by + cz = d$ is also the equation of the plane.

$[2] - [1]$  $\Rightarrow$  $2a - c = 0$  [4]

$[3] + [4]$  $\Rightarrow$  $2c = d$

$\therefore$  $c = \tfrac{1}{2}d$, $a = \tfrac{1}{4}d$ and $b = \tfrac{1}{2}d$

$\therefore$  the equation of the plane is $\tfrac{1}{4}dx + \tfrac{1}{2}dy + \tfrac{1}{2}dz = d$  $\Rightarrow$  $x + 2y + 2z = 4$

# LINES AND PLANES

A line can be either in a plane, or parallel to a plane or it can intersect the plane.

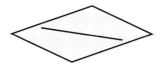  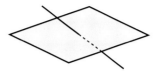

## A line that is in the plane

A line with equation $\mathbf{r} = \mathbf{a} + t\mathbf{b}$ is in a plane $P$ if any two points on the line are also points in the plane. We can take any two values of $t$ to find the position vectors of two points on the line. If these two position vectors also satisfy the equation of the plane, then the line must be in the plane.

## A line that is parallel to the plane

A line with equation $\mathbf{r} = \mathbf{a} + t\mathbf{b}$ is parallel to $\mathbf{b}$. If $\mathbf{n}$ is the vector perpendicular to a plane $P$, and this line is parallel to $P$, then $\mathbf{b}$ is also perpendicular to $\mathbf{n}$. Therefore $\mathbf{b}.\mathbf{n} = 0$

## A line that intersects a plane

A line that is not parallel to a plane must intersect the plane. So the condition for a line to intersect a plane is $\mathbf{b}.\mathbf{n} \neq 0$

---

### Examples 14c cont.

2 Show that the line $L$ whose vector equation is $\mathbf{r} = 2\mathbf{i} - 2\mathbf{j} + 3\mathbf{k} + t(\mathbf{i} - \mathbf{j} + 4\mathbf{k})$ is parallel to the plane $P$ whose vector equation is $\mathbf{r}.(\mathbf{i} + 5\mathbf{j} + \mathbf{k}) = 5$

The line $L$ is parallel to the vector $\mathbf{i} - \mathbf{j} + 4\mathbf{k}$ and the plane $P$ is perpendicular to the vector $\mathbf{i} + 5\mathbf{j} + \mathbf{k}$ so $L$ is parallel to $P$ if and only if $\mathbf{i} - \mathbf{j} + 4\mathbf{k}$ and $\mathbf{i} + 5\mathbf{j} + \mathbf{k}$ are perpendicular.

$L$ is parallel to $P$ if and only if $(\mathbf{i} - \mathbf{j} + 4\mathbf{k}).(\mathbf{i} + 5\mathbf{j} + \mathbf{k}) = 0$

$(\mathbf{i} - \mathbf{j} + 4\mathbf{k}).(\mathbf{i} + 5\mathbf{j} + \mathbf{k}) = 1 - 5 + 4 = 0$

$\therefore$ $L$ is parallel to $P$.

To check that the line is not in the plane, we will find the position vector of a point on the line: using $t = 0$ (any value of $t$ will do but choose an easy one) gives $2\mathbf{i} - 2\mathbf{j} + 3\mathbf{k}$. Substituting this for $\mathbf{r}$ in the equation of the plane gives $(2\mathbf{i} - 2\mathbf{j} + 3\mathbf{k}) . (\mathbf{i} + 5\mathbf{j} + 3\mathbf{k}) = 1$ and $1 \neq 5$ so the line is not in the plane.

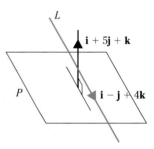

---

## The angle between a line and a plane

A line with equation $\mathbf{r} = \mathbf{a} + t\mathbf{b}$ is parallel $\mathbf{b}$. If $\mathbf{n}$ is the vector perpendicular to a plane $P$, then the angle, $\theta$, between the line and $\mathbf{n}$ can be found from $\mathbf{b}.\mathbf{n}$.

From the diagram, you can see that the angle between the line and the plane is $90° - \theta$

**Examples 14c cont.**

3 The equation of a line is $\mathbf{r} = (1 - s)\mathbf{i} + (2 + 3s)\mathbf{j} + 6s\mathbf{k}$ and the equation of a plane is $\mathbf{r}.(2\mathbf{i} - \mathbf{j} + \mathbf{k}) = 3$. Find

(a) the coordinates of the point where the line cuts the plane

(b) the angle between the line and the plane in degrees correct to 1 decimal place.

(a) The line cuts the plane at the point whose position vector satisfies both the equation of the line and the equation of the plane.

The line cuts the plane where $[(1 - s)\mathbf{i} + (2 + 3s)\mathbf{j} + 6s\mathbf{k}].(2\mathbf{i} - \mathbf{j} + \mathbf{k}) = 3$

$\Rightarrow \quad 2(1 - s) - (2 + 3s) + 6s = 3 \quad \Rightarrow \quad s = 3$

When $s = 3$, $\mathbf{r} = -2\mathbf{i} + 11\mathbf{j} + 18\mathbf{k}$

$\therefore$    the line and the plane meet at the point $(-2, 11, 18)$.

(b) Rearranging the equation of the line as $\mathbf{r} = \mathbf{i} + 2\mathbf{j} + s(-\mathbf{i} + 3\mathbf{j} + 6\mathbf{k})$ shows that the line is parallel to the vector $-\mathbf{i} + 3\mathbf{j} + 6\mathbf{k}$.

The plane is perpendicular to the vector $2\mathbf{i} - \mathbf{j} + \mathbf{k}$.

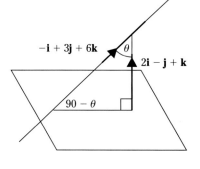

The diagram shows that we can determine the angle between the line and the plane by using the scalar product to find the angle between the line and the normal to the plane.

$(-\mathbf{i} + 3\mathbf{j} + 6\mathbf{k}).(2\mathbf{i} - \mathbf{j} + \mathbf{k})$

$\qquad = |-\mathbf{i} + 3\mathbf{j} + 6\mathbf{k}| \, |2\mathbf{i} - \mathbf{j} + \mathbf{k}| \cos \theta$

$\Rightarrow \quad 1 = \sqrt{46}\sqrt{6} \cos \theta \quad \Rightarrow \quad \theta = 86.54...$

$\therefore$    the angle between the line and the plane is $90° - 86.54...°$

i.e.    $3.5°$ correct to 1 decimal place.

4 The lines $\mathbf{r}_1 = \mathbf{i} - \mathbf{j} + 3\mathbf{k} + \lambda(\mathbf{i} - \mathbf{j} + \mathbf{k})$ and $\mathbf{r}_2 = 2\mathbf{i} + 4\mathbf{j} + 6\mathbf{k} + \mu(2\mathbf{i} + \mathbf{j} + 3\mathbf{k})$ intersect at the point with position vector $-2\mathbf{i} + 2\mathbf{j}$. (This is shown in Examples 14b, question 1.)

Find the Cartesian equation of the plane containing the lines.

To find the equation of a plane we need a point on the plane that is given. We also need a vector, $\mathbf{n}$, that is perpendicular to the plane.

$\mathbf{n} = a\mathbf{i} + b\mathbf{j} + c\mathbf{k}$ is a vector that is perpendicular to the plane, so $\mathbf{n}$ is also perpendicular to any line contained in the plane.

$\therefore \qquad (a\mathbf{i} + b\mathbf{j} + c\mathbf{k}).(\mathbf{i} - \mathbf{j} + \mathbf{k}) = 0$

and $\qquad (a\mathbf{i} + b\mathbf{j} + c\mathbf{k}).(2\mathbf{i} + \mathbf{j} + 3\mathbf{k}) = 0$

The scalar product of $\mathbf{n}$ and any vector parallel to the plane is zero. We use the equations of the lines to write down two vectors that are parallel to the plane.

So $\qquad\qquad\qquad\qquad a - b + c = 0 \qquad\qquad$ [1]    We can use these equations to find $a$ and $b$ in terms of $c$.

and $\qquad\qquad\qquad\qquad 2a + b + 3c = 0 \qquad\quad$ [2]

[1] gives $b = a + c$ and [2] gives $b = -2a - 3c$

$\therefore \quad a + c = -2a - 3c$ so $a = -\frac{4}{3}c$ and $b = -\frac{1}{3}c$

$\therefore \quad -\frac{4}{3}c\mathbf{i} - \frac{1}{3}c\mathbf{j} + c\mathbf{k}$ is perpendicular to the plane and so is $4\mathbf{i} + \mathbf{j} - 3\mathbf{k}$    Multiply by $-\frac{3}{c}$

The equation of the plane is $\{(x\mathbf{i} + y\mathbf{j} + z\mathbf{k} - (-2\mathbf{i} + 2\mathbf{j})\}.(4\mathbf{i} + \mathbf{j} - 3\mathbf{k}) = 0$

$\Rightarrow \quad \{(x + 2)\mathbf{i} + (y - 2)\mathbf{j} + z\}.(4\mathbf{i} + \mathbf{j} - 3\mathbf{k}) = 0$       Using $(\mathbf{r} - \overrightarrow{OA}).\mathbf{n} = 0$

$\Rightarrow \quad 4x + 8 + y - 2 - 3z = 0$

i.e.    $4x + y - 3z = -6$

## The distance of a point from a plane

In the diagram, $\mathbf{a}$ is the position vector of the point $A$, $\hat{\mathbf{n}}$ is a unit vector perpendicular to the plane and $d$ is the distance of the plane from the origin.

The distance of $A$ from the plane is $OB - d$

$OB = OA \cos \theta = \hat{\mathbf{n}}.\mathbf{a}$

Therefore the distance of $A$ from the plane is $\hat{\mathbf{n}}.\mathbf{a} - d$

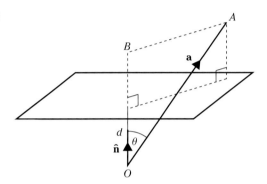

### Examples 14c cont.

5   Find the distance of the point $A(1, 5, 0)$ from the plane $P$ with equation $3x + y - z = 11$

$OB = \hat{\mathbf{n}}.\overrightarrow{OA}$

the distance of $A$ from $P$

$$= \frac{1}{\sqrt{11}}(3\mathbf{i} + \mathbf{j} - \mathbf{k}).(\mathbf{i} + 5\mathbf{j}) - \frac{11}{\sqrt{11}}$$

$$= \frac{8}{\sqrt{11}} - \frac{11}{\sqrt{11}} = -\frac{3}{\sqrt{11}}$$

Because this is negative it shows that $O$ and $A$ are on the same side of $P$.

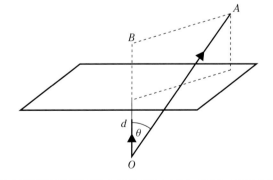

## The line of intersection of two planes

Two planes are either parallel (in which case the perpendicular vectors to the planes will be parallel) or they intersect.

When they intersect, the equation of the line between the planes can be found by using the Cartesian equations of the planes. This is shown in the next worked example.

## Examples 14c cont.

**6** The planes with equations $x - 2y + z = 2$ and $2x - 3y - z = 4$ intersect in the line $l$. Find the equation of $l$.

The coordinates of any point $(a, b, c)$ on $l$ satisfy the equations of both planes.

Therefore   $\begin{cases} a - 2b + c = 2 & [1] \\ 2a - 3b - c = 4 & [2] \end{cases}$   $\Rightarrow$   $3a - 5b = 6$   [3]

Let $b = 3$, then [3] gives $a = 7$ and [1] gives $c = 1$

Let $b = 6$, then [3] gives $a = 12$ and [1] gives $c = 2$

We can choose any value for one of the letters. The values chosen here do not give fractions for the other two letters.

$\therefore$    $A(7, 3, 1)$ and $B(12, 6, 2)$ are on $l$

$\therefore$    an equation of $l$ is $\mathbf{r} = 7\mathbf{i} + 3\mathbf{j} + \mathbf{k} + s(5\mathbf{i} + 3\mathbf{j} + \mathbf{k})$

Using $\overrightarrow{OA}$ as a point on $l$ and $\overrightarrow{BA}$ parallel to $l$.

**7** Find the angle between the planes with equations $x - 2y + z = 2$ and $2x - 3y - z = 4$

The angle $\theta$ between the planes is equal to the angle between the normals to the planes, i.e. the angle between

$$\mathbf{i} - 2\mathbf{j} + \mathbf{k} \text{ and } 2\mathbf{i} - 3\mathbf{j} - \mathbf{k}$$

$(\mathbf{i} - 2\mathbf{j} + \mathbf{k}).(2\mathbf{i} - 3\mathbf{j} - \mathbf{k}) = \sqrt{6} \times \sqrt{14} \cos \theta$

$\Rightarrow$    $\cos \theta = \dfrac{2 + 6 - 1}{\sqrt{6} \times \sqrt{14}}$

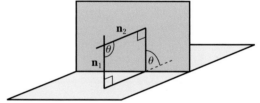

Therefore $\theta = 40.2°$

## Exercise 14c

**1** Write down the Cartesian equations of these planes.

(a)   $\mathbf{r}.(\mathbf{i} + \mathbf{j} - \mathbf{k}) = 2$

(b)   $\mathbf{r}.(2\mathbf{i} + 3\mathbf{j} - 4\mathbf{k}) = 1$

**2** Write down the vector equations of these planes.

(a)   $3x - 2y + z = 5$

(b)   $5x - 3y - 4z = 7$

**3** Write down a vector that is perpendicular to each plane given in questions **1** and **2**.

**4** Find the distance from the origin of each plane given in questions **1** and **2**.

**5** Two planes $P_1$ and $P_2$ have vector equations $\mathbf{r}.(2\mathbf{i} + \mathbf{j} - 2\mathbf{k}) = 3$ and $\mathbf{r}.(2\mathbf{i} + \mathbf{j} - 2\mathbf{k}) = 9$

Explain why $P_1$ and $P_2$ are parallel and hence find the distance between them.

**6** Find the vector equation of the line through the origin that is perpendicular to the plane $\mathbf{r}.(\mathbf{i} - 2\mathbf{j} + \mathbf{k}) = 3$

**7** Find the vector equation of the line through the point $(2, 1, 1)$ that is perpendicular to the plane $\mathbf{r}.(\mathbf{i} + 2\mathbf{j} - 3\mathbf{k}) = 6$

**8** A plane goes through the three points whose position vectors are $\mathbf{a}$, $\mathbf{b}$ and $\mathbf{c}$ where

(a)   $= \mathbf{i} + \mathbf{j} + 2\mathbf{k}$        (b)   $= 2\mathbf{i} - \mathbf{j} + 3\mathbf{k}$

(c)   $= -\mathbf{i} + 2\mathbf{j} - 2\mathbf{k}$

Find the vector equation of this plane in scalar product form and hence find the distance of the plane from the origin.

9   Find the vector equation of the plane that contains the points $A(0, 1, 1)$, $B(-1, 2, 1)$ and $C(2, 0, 2)$.

10  Find the vector equation of the plane that contains the lines

$\mathbf{r} = -3\mathbf{i} - 2\mathbf{j} + t(\mathbf{i} - 2\mathbf{j} + \mathbf{k})$ and

$\mathbf{r} = \mathbf{i} - 11\mathbf{j} + 4\mathbf{k} + s(2\mathbf{i} - \mathbf{j} + 2\mathbf{k})$

11  Find the point of intersection of the line $\mathbf{r} = (\mathbf{i} + \mathbf{j} - 2\mathbf{k}) + \lambda(\mathbf{i} - \mathbf{j} + \mathbf{k})$ and the plane $\mathbf{r}.(\mathbf{i} + 2\mathbf{j} - \mathbf{k}) = 2$

12

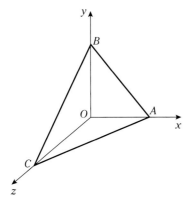

$OABC$ is a tetrahedron. The coordinates of $A$, $B$ and $C$ are $(2, 0, 0)$, $(0, 3, 0)$ and $(0, 0, 4)$ respectively. Find

(a)  a Cartesian equation for the plane $ABC$

(b)  the distance of the plane $ABC$ from $O$

(c)  the angle between the edge $OB$ and the plane $ABC$.

13  A line passes through the point with position vector $2\mathbf{i} + \mathbf{j} - 4\mathbf{k}$.

The line is parallel to the vector $\mathbf{i} - \mathbf{j} + 3\mathbf{k}$.

(a)  Write down the equation of this line.

The equation of a plane is $2x - 5y + z = 8$

(b)  Find the position vector of the point where the line intersects the plane.

(c)  Find the acute angle between the line and the plane.

14  (a)  Find the equation of the line of intersection of the planes $x + y - z = 5$ and $2x - y + z = 1$

(b)  Find the angle between the planes in part (a).

15  Repeat question 14 for the planes in question 1.

16  The position vectors of three points are $\mathbf{i} + \mathbf{j} - 4\mathbf{k}$, $2\mathbf{i} - 3\mathbf{j} - \mathbf{k}$ and $2\mathbf{i} - \mathbf{j} - 2\mathbf{k}$.

Find the equation of the plane containing these three points.

# 15 Complex numbers

## After studying this chapter you should be able to

- understand the idea of a complex number, recall the meaning of the terms real part, imaginary part, modulus, argument, conjugate, and use the fact that two complex numbers are equal if and only if both real and imaginary parts are equal
- carry out operations of addition, subtraction, multiplication and division of two complex numbers expressed in Cartesian form $x + iy$
- use the result that, for a polynomial equation with real coefficients, any non-real roots occur in conjugate pairs
- represent complex numbers geometrically by means of an Argand diagram
- carry out operations of multiplication and division of two complex numbers expressed in polar form $r(\cos \theta + i \sin \theta) \equiv re^{i\theta}$
- find the two square roots of a complex number
- understand in simple terms the geometrical effects of conjugating a complex number and of adding, subtracting, multiplying and dividing two complex numbers
- illustrate simple equations and inequalities involving complex numbers by means of loci in an Argand diagram, e.g. $|z - a| < k$, $|z - a| = |z - b|$, $\arg(z - a) = \alpha$

## IMAGINARY NUMBERS

A real number is any number we can represent as a point on a number line.

When a real number is squared, the result is a positive number, e.g. $3^2 = 9$ and $(-2)^2 = 4$

An equation of the form $x^2 = -1$ does not have real roots because there is no real number that when squared gives $-1$.

To work with equations whose roots are not real, we need numbers whose squares are negative.

These numbers are called imaginary numbers, for example $\sqrt{-4}, \sqrt{-9}, \sqrt{-11}$.

We use i to denote $\sqrt{-1}$.

All other imaginary numbers can be written in terms of i.

For example, $\sqrt{-4} = \sqrt{4 \times -1} = \sqrt{4} \times \sqrt{-1} = 2i$

and $\sqrt{-11} = \sqrt{11 \times -1} = \sqrt{11} \times \sqrt{-1} = (\sqrt{11})i$

Imaginary numbers can be added, subtracted, multiplied and divided.

For example, $2i + 5i = 7i$

$\sqrt{7}i - i = (\sqrt{7} - 1)i$

$2i \times 5i = 10i^2$

But $i^2 = -1$ since $i = \sqrt{-1}$

Therefore $2i \times 5i = -10$

Also $6i \div 3i = 2$

Powers of i can be simplified,

e.g. $i^3 = (i^2)(i) = -i$, $i^4 = (i^2)^2 = (-1)^2 = 1$ and $i^{-1} = \dfrac{1}{i} = \dfrac{i}{i^2} = \dfrac{i}{-1} = -i$

# COMPLEX NUMBERS

When a real number and an imaginary number are added or subtracted, the expression formed is called a complex number, e.g. $2 + 3i$, $4 - 7i$, $-1 + 4i$. Any complex number can be written as $a + bi$ where $a$ and $b$ can have any real value including zero.

If     $a = 0$     the number is $bi$, i.e. an imaginary number.

If     $b = 0$     the number is $a$, i.e. a real number.

Two complex numbers are equal if and only if the real parts are equal and the imaginary parts are equal, i.e. if $a + bi = c + di$, then $a = c$ and $b = d$.

## OPERATIONS ON COMPLEX NUMBERS

### Addition and subtraction

Real terms and imaginary terms are added and subtracted separately,

e.g. $\qquad (2 + 3i) + (4 - i) = (2 + 4) + (3i - i)$
$$= 6 + 2i$$

and $\qquad (4 - 2i) - (3 + 5i) = (4 - 3) - (2i + 5i)$
$$= 1 - 7i$$

### Multiplication

We use the same rule for multiplying two complex numbers as we use to find $(a + b)(c + d)$,

e.g. $\qquad (2 + 3i)(4 - i) = 8 - 2i + 12i - 3i^2$
$$= 8 + 10i - 3(-1)$$
$$= 11 + 10i$$

and $\qquad (2 + 3i)(2 - 3i) = 4 - 6i + 6i - 9i^2$
$$= 4 + 9$$
$$= 13$$

The product in the last example is a real number. This is true for any product of the form

$\qquad (a + ib)(a - ib) \qquad$ because $\qquad (a + ib)(a - ib) \equiv a^2 - iab + iab + i^2b^2 \equiv a^2 + b^2$

Complex numbers such as $a + ib$ and $a - ib$ are called conjugate and each is the conjugate of the other.

So $4 + 5i$ and $4 - 5i$ are conjugate complex numbers and $4 + 5i$ is the conjugate of $4 - 5i$.

When $z = a + ib$, its conjugate $a - ib$ is denoted by $z^*$.

### Division

Dividing one complex number by another can be done by using conjugate complex numbers, for example

$$\frac{2 + 9i}{5 - 2i} = \frac{(2 + 9i)(5 + 2i)}{(5 - 2i)(5 + 2i)}$$

(multiplying numerator and denominator by the conjugate of the denominator)

$$= \frac{10 + 49i + 18i^2}{25 - 4i^2}$$
$$= \frac{-8 + 49i}{29}$$
$$= -\frac{8}{29} + \frac{49}{29}i$$

(The real term is given first even when it is negative.)

The process we have used changes the denominator from a complex number to a real number. We call this realising the denominator.

## Example 15a

Find the values of $x$ and $y$ for which $(x + iy)(2 - 3i) = 8 + i$

$$(x + iy)(2 - 3i) = 2x + 3y + i(2y - 3x)$$

$\therefore$    $(x + iy)(2 - 3i) = 8 + i \Leftrightarrow 2x + 3y + i(2y - 3x) = 8 + i$

Equating real and imaginary parts gives

$$2x + 3y = 8 \qquad [1]$$

and    $2y - 3x = 1 \qquad [2]$

Solving these equations simultaneously gives

$[1] \times 3 + [2] \times 2$    gives    $13y = 26$    so    $y = 2$

From [1]    $2x + 6 = 8$    so    $x = 1$

## Exercise 15a

**1** Simplify: $i^7$, $i^{-3}$, $i^9$, $i^{-5}$, $i^{4n}$, $i^{4n+1}$.

**2** Add the following pairs of complex numbers:

(a)   $3 + 5i$ and $7 - i$    (b)   $4 - i$ and $3 + 3i$

(c)   $2 + 7i$ and $4 - 9i$    (d)   $a - bi$ and $c + di$

**3** Subtract the second number from the first in each part in question **2**.

**4** Simplify:

(a)   $(2 + i)(3 - 4i)$    (b)   $(5 + 4i)(7 - i)$

(c)   $(3 - i)(4 - i)$    (d)   $(3 + 4i)(3 - 4i)$

(e)   $(2 - i)^2$    (f)   $(1 + i)^3$

(g)   $i(3 + 4i)$    (h)   $(x + iy)(x - iy)$

(i)   $i(1 + i)(2 + i)$    (j)   $(a + bi)^2$

**5** Express each of the following fractions in the form $a + bi$.

(a)   $\dfrac{2}{1 - i}$    (b)   $\dfrac{3 + i}{4 - 3i}$

(c)   $\dfrac{4i}{4 + i}$    (d)   $\dfrac{1 + i}{1 - i}$

(e)   $\dfrac{7 - i}{1 + 7i}$    (f)   $\dfrac{x + iy}{x - iy}$

(g)   $\dfrac{3 + i}{i}$    (h)   $\dfrac{-2 + 3i}{-i}$

**6** Solve the following equations for $x$ and $y$.

(a)   $x + iy = (3 + i)(2 - 3i)$

(b)   $\dfrac{2 + 5i}{1 - i} = x + iy$

(c)   $3 + 4i = (x + iy)(1 + i)$

(d)   $x + iy = 2$

(e)   $x + iy = (3 + 2i)(3 - 2i)$

(f)   $x + iy = (4 + i)^2$

(g)   $\dfrac{x + iy}{2 + i} = 5 - i$

(h)   $(x + iy)^2 = 3 + 4i$

**7** Find the real and imaginary parts of:

(a)   $(2 - i)(3 + i)$    (b)   $(1 + i)^3$

(c)   $\dfrac{3 + 2i}{4 - i}$    (d)   $\dfrac{2}{3 + i} + \dfrac{3}{2 + i}$

(e)   $\dfrac{1}{x + iy} - \dfrac{1}{x - iy}$

(f)   $\left( \cos \dfrac{\pi}{3} + i \sin \dfrac{\pi}{3} \right)^3$

(g)   $\left( \cos \dfrac{\pi}{6} + i \sin \dfrac{\pi}{6} \right)^2$

## The two square roots of a complex number

If $x + iy$ is a square root of the complex number $a + ib$, then $(x + iy)^2 = a + ib$

Expanding the left-hand side gives $\qquad\qquad x^2 - y^2 + 2ixy = a + ib$

This gives a pair of simultaneous equations which we can solve to find values for $x$ and $y$.

The equations are quadratic, so there will be two values for $x$ and $y$. Therefore the complex number has two square roots.

---

### Example 15b

Find the square roots of $-1 + 2\sqrt{2}\,i$

If $(x + iy)$ is a square root then

$$(x + iy)^2 = x^2 + 2ixy + (iy)^2 = (x^2 - y^2) + 2ixy$$

Therefore $(x^2 - y^2) + 2ixy = -1 + 2\sqrt{2}\,i$ if and only if $x^2 - y^2 = -1$ and $2xy = 2\sqrt{2}$

Two complex numbers are equal if and only if the real parts are equal and the imaginary parts are equal.

Solving these two equations simultaneously gives $x = \dfrac{\sqrt{2}}{y}$ so

$$\left(\dfrac{\sqrt{2}}{y}\right)^2 - y^2 = -1 \Rightarrow 2 - y^4 = -y^2$$

i.e. $\quad y^4 - y^2 - 2 = 0 \Rightarrow (y^2 - 2)(y^2 + 1) = 0$

so $\quad y^2 = 2$

$y$ is a real number so a negative value of $y^2$ would not give a valid value of $y$.

$\therefore \qquad y = +\sqrt{2} \qquad$ or $\qquad -\sqrt{2}$

$\qquad\qquad x = 1 \qquad\qquad\qquad -1$

The square roots are $(1 + \sqrt{2}i)$ and $-(1 + \sqrt{2}i)$

---

### Exercise 15b

Find the square roots of each complex number.

**1** $3 - 4i$

**2** $2i$

**3** $3 + 4i$

**4** $-1 + 2\sqrt{6}i$

**5** $4 - 4i$

## COMPLEX ROOTS OF QUADRATIC EQUATIONS

The roots of the quadratic equation $x^2 + 2x + 2 = 0$ can be found using the formula

$x = \dfrac{-b \pm \sqrt{b^2 - 4ac}}{2a} \Rightarrow x = \dfrac{-2 \pm \sqrt{-4}}{2}$ so the roots of the given equation are the complex numbers

$-1 + i$ and $-1 - i$. Also, the roots are conjugate complex numbers.

When $\qquad ax^2 + bx + c = 0 \qquad\qquad\qquad$ [1]

and $\qquad\qquad b^2 - 4ac < 0$

then $\qquad\qquad x = \dfrac{-b \pm \sqrt{b^2 - 4ac}}{2a}$

$\Rightarrow \qquad\qquad x = \dfrac{-b}{2a} \pm \dfrac{i\sqrt{4ac - b^2}}{2a}$

Now using $\qquad p = \dfrac{-b}{2a} \qquad$ and $\qquad q = \dfrac{\sqrt{4ac - b^2}}{2a}$

the roots of equation [1] are

$\qquad p + qi \qquad$ and $\qquad p - qi$

and these are conjugate complex numbers.

This means that

**the complex roots of a quadratic equation are a pair of conjugate complex numbers.**

## The roots of a cubic equation

Sketches of the possible curves of $y = ax^3 + bx^2 + cx + d$ show that the equation $ax^3 + bx^2 + cx + d = 0$ has either one real root or three real roots (which may include a repeated root).

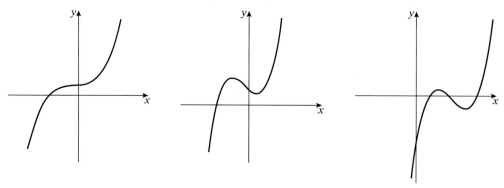

Therefore the equation $ax^3 + bx^2 + cx + d = 0$ can always be written as the product of a linear factor and a quadratic factor,

i.e. as $(x - p)(ax^2 + bx + c) = 0$ where $p$ is a real root of the equation.

$\therefore \quad$ if $ax^3 + bx^2 + cx + d = 0$ has only one real root, it also has a pair of conjugate complex roots.

Any polynomial equation with real coefficients can be expressed as a product of quadratic factors and possibly linear factors. This means that any complex roots occur as conjugate pairs.

---

### Examples 15c

1 One root of the equation $x^2 + px + q = 0$ is $2 - 3i$. Find the values of $p$ and $q$.

One root is $\alpha = 2 - 3i$ so the other is $\beta = 2 + 3i$.

Then $\qquad \alpha + \beta = 4$

and $\qquad\quad \alpha\beta = (2 - 3i)(2 + 3i)$

$\qquad\qquad\quad = 13$

Any quadratic equation can be written in the form

$$x^2 - (\text{sum of roots})x + (\text{product of roots}) = 0$$

So the equation with roots $2 \pm 3i$ is

$$x^2 - 4x + 13 = 0$$

i.e.    $p = 4$    and    $q = 13$

2  The complex number $3 - 2i$ is a root of the equation $x^3 - 7x^2 + kx - 13 = 0$
Find the value of $k$.

The complex roots occur as a conjugate pair so $3 - 2i$ and $3 + 2i$ are roots.

$\therefore$    $(x - (3 - 2i))$ and $(x - (3 + 2i))$ are factors.

The quadratic factor is $(x - (3 - 2i))(x - (3 + 2i)) = x^2 - 6x + 13$

$\therefore$    $x^3 - 7x^2 + kx - 13 = (x - a)(x^2 - 6x + 13)$

$\Rightarrow$                    $-13 = -13a$    so    $a = 1$

and    $kx = 13x + 6ax$    so    $k = 13 + 6 \Rightarrow k = 19$

## Exercise 15c

1  Solve the following equations.

(a)  $x^2 + x + 1 = 0$

(b)  $2x^2 + 7x + 1 = 0$

(c)  $x^2 + 9 = 0$

(d)  $x^2 + x + 3 = 0$

(e)  $x^4 - 1 = 0$

2  Form the equation whose roots are

(a)  $i, -i$                (b)  $2 + i, 2 - i$

(c)  $1 - 3i, 1 + 3i$        (d)  $1 + i, 1 - i, 2$

3  The complex number $2 - i$ is one root of the
equation $x^3 - 5x^2 + ax - 5$

Find the value of $a$.

# THE ARGAND DIAGRAM

The complex number $a + ib$ can be represented by the ordered pair $\begin{pmatrix} a \\ ib \end{pmatrix}$. By using $a$ and $ib$ as the
coordinates of a point $A$, we can use the vector $\overrightarrow{OA}$ as a visual representation of the complex number
$a + ib$

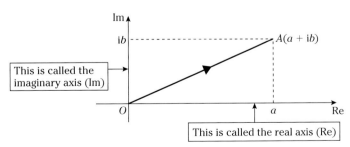

This diagram is called an Argand diagram.

Any complex number $x + iy$ is represented by $\overrightarrow{OP}$ where $P$ is the point $x + iy$

## A COMPLEX NUMBER AS A VECTOR

On an Argand diagram, the complex number $5 + 3i$ can be represented by the line $\overrightarrow{OA}$ where $A$ is the point $5 + 3i$. Any other line with the same length and direction (e.g. $\overrightarrow{BC}$ or $\overrightarrow{DE}$) can also be used.

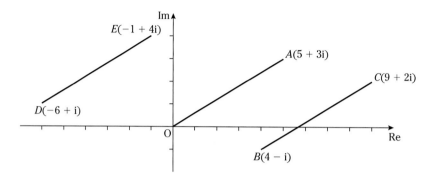

So a complex number can be represented by a displacement vector. It can also be represented by a position vector, when it is sometimes represented by the point $A$.

We use $z$ to denote a complex number, e.g. $z = x + iy$, $z_1 = 5 + 3i$

The line representing $z$ on an Argand diagram must have an arrow to show the direction of the line.

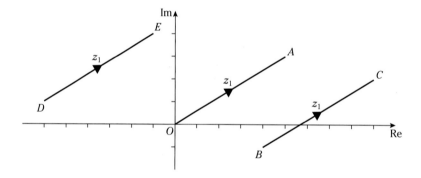

## GRAPHICAL REPRESENTATION OF ADDITION AND SUBTRACTION

Two complex numbers $z_1$ and $z_2$ are represented on an Argand diagram by $\overrightarrow{OA}$ and $\overrightarrow{OB}$.

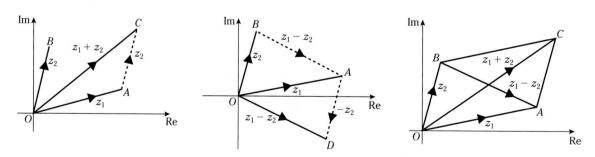

Using vector addition and subtraction shows that the two diagonals of the parallelogram $OACB$ represent the sum and difference of $z_1$ and $z_2$

## Conjugate complex numbers

When $z = x + iy$ and its conjugate $z^* = x - iy$ are represented on an Argand diagram, we see that $z^*$ is the reflection of $z$ in the real axis.

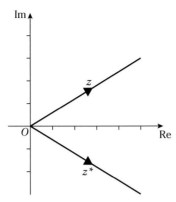

# MODULUS AND ARGUMENT

The point $A$ can also be located using polar coordinates $(r, \theta)$ where $r$ is the length of $\overrightarrow{OA}$ and $\theta$ is the angle between the positive $x$-axis and $\overrightarrow{OA}$.

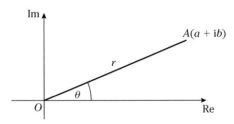

The length of $OA$ is called the *modulus* of the complex number $a + ib$ and is written $|a + ib|$ so that

$$|a + ib| = r = \sqrt{a^2 + b^2}$$

The angle $\theta$ is called the *argument* of $a + ib$ and is written $\arg(a + ib)$,

Therefore $\arg(a + ib) = \theta$ where $\tan \theta = \dfrac{b}{a}$ and $-\pi \leqslant \theta \leqslant \pi$

To find the value of $\theta$ for a particular complex number, we draw it on an Argand diagram:

For example, the diagrams below represent the complex numbers $4 + 3i$, $-4 + 3i$, $-4 - 3i$ and $4 - 3i$

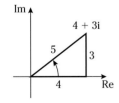 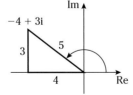 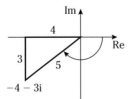 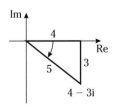

$4 + 3i$ is in the first quadrant, so $\arg(4 + 3i)$ is positive and acute.

Therefore $\tan \theta = \frac{3}{4}$ so $\theta = 0.644$ rad

$-4 + 3i$ is in the second quadrant so $\arg(-4 + 3i)$ is positive and obtuse.

Therefore $\tan \theta = -\frac{3}{4}$ so $\theta = 2.498$ rad

$-4 - 3i$ is in the third quadrant so $\arg(-4 - 3i)$ is negative and obtuse so its value is $-2.498$ rad.

$4 - 3i$ is in the fourth quadrant so $\arg(4 - 3i)$ is negative and acute so its value is $-0.644$ rad.

# THE POLAR COORDINATE FORM OF A COMPLEX NUMBER

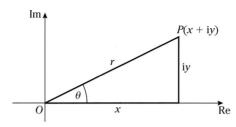

The diagram shows that $\quad x \equiv r \cos \theta$

and $\qquad\qquad\qquad\qquad y \equiv r \sin \theta$

Therefore $\qquad\qquad\quad x + iy \equiv r \cos \theta + ir \sin \theta$

i.e. $\qquad\qquad\qquad\quad x + iy \equiv r(\cos \theta + i \sin \theta)$

$r\,(\cos \theta + i \sin \theta)$ is called the polar form of a complex number.

When a complex number is given as $x + iy$ it can be converted to polar form by finding the modulus and argument.

For example, for $1 - i$,

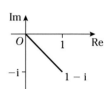

$$r = \sqrt{1 + 1} = \sqrt{2} \qquad \text{and} \qquad \theta = -\frac{\pi}{4}$$

Therefore $\qquad 1 - i = \sqrt{2}\left( \cos\left(-\frac{\pi}{4}\right) + i \sin\left(-\frac{\pi}{4}\right)\right)$

A complex number in polar form can easily be expressed as $x + iy$.

For example, $\qquad 4\left( \cos\frac{2\pi}{3} + i \sin\frac{2\pi}{3} \right) = 4\left( -\frac{1}{2} + i\frac{\sqrt{3}}{2} \right)$

$$= -2 + 2i\sqrt{3}$$

---

## Examples 15d

1  Given that $z_1 = 3 - i$ and $z_2 = -2 + 5i$, represent on an Argand diagram the complex numbers $z_1, z_2, z_1 + z_2, z_1 - z_2$. Find the modulus and argument of $z_1 + z_2$ and $z_1 - z_2$

$\overrightarrow{OA}$ represents $z_1$

$\overrightarrow{OB}$ represents $z_2$

$\overrightarrow{OC}$ represents $z_1 + z_2 = 1 + 4i$

$\overrightarrow{OD}$ (equal and parallel to $\overrightarrow{BA}$)
      represents $z_1 - z_2 = 5 - 6i$

$|z_1 + z_2| = \sqrt{1^2 + 4^2} = \sqrt{17}$ (length of $OC$)

$|z_1 - z_2| = \sqrt{5^2 + (-6)^2} = \sqrt{61}$ (length of $OD$)

$\arg(z_1 + z_2) =$ the angle whose tan is $\left(\frac{4}{1}\right)$
            $= 1.33\,\text{rad}$

$\arg(z_1 - z_2) =$ the angle whose tan is $\left(-\frac{6}{5}\right)$
            $= -0.88\,\text{rad}$

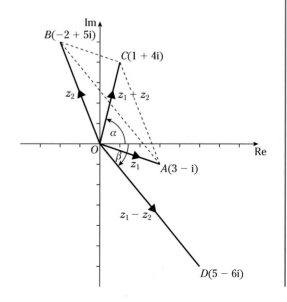

**2** Find the modulus and argument of $\dfrac{7 - i}{3 - 4i}$

First express $\dfrac{7 - i}{3 - 4i}$ in the form $a + ib$

$$\dfrac{7 - i}{3 - 4i} = \dfrac{(7 - i)(3 + 4i)}{(3 - 4i)(3 + 4i)}$$

$$= \dfrac{25 + 25i}{25} = 1 + i$$

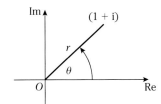

Therefore $|1 + i| = \sqrt{1^2 + 1^2} = \sqrt{2}$

and $\arg(1 + i)$ is a positive acute angle, $\theta$, where $\tan\theta = 1$, so $\theta = \dfrac{\pi}{4}$

**3** Express in the form $r(\cos\theta + i\sin\theta)$:

  (a)  $1 - i\sqrt{3}$                 (b)  2                     (c)  $-5i$

(a)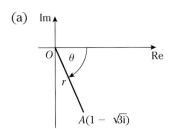

For $1 - i\sqrt{3}$

$$r = \sqrt{\left[1^2 + \left(-\sqrt{3}\right)^2\right]} = 2$$

$$\tan\theta = -\sqrt{3}$$

so $$\theta = -\dfrac{\pi}{3}$$

$\therefore$ $$1 - i\sqrt{3} = 2\left[\cos\left(-\dfrac{\pi}{3}\right) + i\sin\left(-\dfrac{\pi}{3}\right)\right]$$

(b)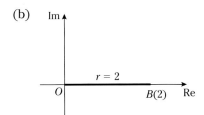

For 2

$$r = 2 \quad \text{and}$$

$$\theta = 0$$

$\therefore$ $$2 = 2(\cos 0 + i\sin 0)$$

(c)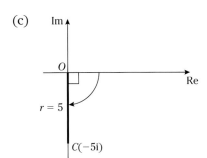

For $-5i$

$$r = 5 \quad \text{and}$$

$$\theta = -\dfrac{\pi}{2}$$

$\therefore$ $$-5i = 5\left\{\cos\left(-\dfrac{\pi}{2}\right) + i\sin\left(-\dfrac{\pi}{2}\right)\right\}$$

**4** $z = 2\left(\cos\frac{\pi}{4} + i\sin\frac{\pi}{4}\right)$. Illustrate $z$ and $z^*$ on an Argand diagram.

$$z^* = 2\left(\cos\frac{\pi}{4} - i\sin\frac{\pi}{4}\right) = 2\left(\cos\left(-\frac{\pi}{4}\right) + i\sin\left(-\frac{\pi}{4}\right)\right)$$

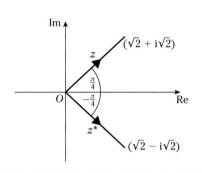

## Exercise 15d

**1** Represent the following complex numbers by lines on Argand diagrams.

Find the modulus and argument of each complex number.

(a) $3 - 2i$        (b) $-4 + i$

(c) $-3 - 4i$        (d) $5 + 12i$

(e) $1 - i$        (f) $-1 + i$

(g) $4$        (h) $-2i$

(i) $a + bi$        (j) $1 + i$

(k) $i(1 + i)$        (l) $i^2(1 + i)$

(m) $i^3(1 + i)$        (n) $(3 + i)(4 + i)$

(o) $2\left(\cos\frac{\pi}{3} + i\sin\frac{\pi}{3}\right)$

(p) $\cos\frac{3\pi}{4} + i\sin\frac{3\pi}{4}$

(q) $3\left[\cos\left(-\frac{5\pi}{6}\right) + i\sin\left(-\frac{5\pi}{6}\right)\right]$

**2** $z_1 = 3 - i$, $z_2 = 1 + 4i$, $z_3 = -4 + i$, $z_4 = -2 - 5i$. Represent the following by lines on Argand diagrams, showing the direction of each line by an arrow.

(a) $z_1 + z_2$        (b) $z_2 - z_3$

(c) $z_1 - z_3$        (d) $z_2 + z_4$

(e) $z_4 - z_1$        (f) $z_3 - z_4$

(g) $z_1$        (h) $z_4$

(i) $z_2 - z_1$        (j) $z_1 + z_3$

**3** Express in the form $r(\cos\theta + i\sin\theta)$:

(a) $1 + i$        (b) $\sqrt{3} - i$

(c) $-3 - 4i$        (d) $-5 + 12i$

(e) $2 - i$        (f) $6$

(g) $-3$        (h) $4i$

(i) $-3 - i\sqrt{3}$        (j) $24 + 7i$

**4** Express in the form $x + iy$ the complex number:

(a) $2\left(\cos\frac{\pi}{6} + i\sin\frac{\pi}{6}\right)$

(b) $3\left(\cos\left(-\frac{\pi}{4}\right) + i\sin\left(-\frac{\pi}{4}\right)\right)$

(c) $\cos\frac{2\pi}{3} + i\sin\frac{2\pi}{3}$

(d) $\cos\left(-\frac{3\pi}{4}\right) + i\sin\left(-\frac{3\pi}{4}\right)$

(e) $3$

(f) $2(\cos\pi + i\sin\pi)$

(g) $4\left(\cos\left(-\frac{\pi}{6}\right) + i\sin\left(-\frac{\pi}{6}\right)\right)$

(h) $\cos\pi + i\sin\pi$

(i) $3\left(\cos\left(-\frac{\pi}{2}\right) + i\sin\left(-\frac{\pi}{2}\right)\right)$

(j) $\cos\left(-\frac{2\pi}{3}\right) + i\sin\left(-\frac{2\pi}{3}\right)$

**5** By using $z_1 = x_1 + iy_1$, $z_2 = x_2 + iy_2$ show on an Argand diagram the position of the point representing:

(a) $\frac{1}{2}(z_1 + z_2)$        (b) $\frac{1}{3}(2z_1 + z_2)$

## The exponential form $z = re^{i\theta}$

Euler's formula states that $e^{i\theta} = \cos\theta + i\sin\theta$, therefore the polar form of a complex number, i.e. $z = r(\cos\theta + i\sin\theta)$, can also be written as $z = re^{i\theta}$

Euler's formula can be proved, but the mathematics needed is not covered in this book.

When $\qquad z_1 = r_1 e^{i\theta_1} \qquad$ and $\qquad z_2 = r_2 e^{i\theta_2}$

then $\qquad z_1 \times z_2 = r_1 e^{i\theta_1} \times r_2 e^{i\theta_2}$

$$= r_1 r_2 \times e^{i\theta_1} e^{i\theta_2} = r_1 r_2 e^{i(\theta_1 + \theta_2)}$$

Therefore $\qquad |z_1 z_2| = |z_1||z_2| \qquad$ and $\qquad \arg(z_1 z_2) = \arg z_1 + \arg z_2$

Also $\qquad \dfrac{z_1}{z_2} = \dfrac{r_1 e^{i\theta_1}}{r_2 e^{i\theta_2}} = \dfrac{r_1}{r_2} \times \dfrac{e^{i\theta_1}}{e^{i\theta_2}} = \dfrac{r_1}{r_2} e^{i(\theta_1 - \theta_2)}$

Therefore $\qquad \left|\dfrac{z_1}{z_2}\right| = \dfrac{|z_1|}{|z_2|} \qquad$ and $\qquad \arg\left(\dfrac{z_1}{z_2}\right) = \arg z_1 - \arg z_2$

## Geometric representation of products and quotients

When $z_1 = 2$, the modulus is 2 and the argument is 0,

So when $\qquad z_2 = r_2 e^{i\theta_2} \qquad z_1 z_2 = 2r_2 e^{i\theta_2}$

Therefore on an Argand diagram, $2z$ is represented by an enlargement of $z$ by a factor of 2.

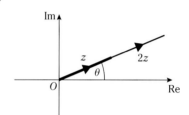

When $z = i$, the modulus is 1 and the argument is $\dfrac{\pi}{2}$, so $i = e^{\frac{i\pi}{2}}$

Therefore when $z = re^{i\theta}$, $iz = e^{\frac{i\pi}{2}} re^{i\theta} = re^{i\left(\theta + \frac{\pi}{2}\right)}$

So, on the Argand diagram, $iz$ is represented by rotating $z$ by $\dfrac{\pi}{2}$ about $O$.

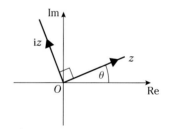

When $z_1 = r_1 e^{i\theta_1} \qquad$ and $\qquad z_2 = r_2 e^{i\theta_2}$

$$z_1 z_2 = r_1 r_2 e^{i(\theta_1 + \theta_2)}$$

On an Argand diagram, you can see that when $z_1$ is multiplied by $z_2$, $z_1$ is enlarged by a scale factor $|z_2|$ and rotated by an angle $\theta_2$.

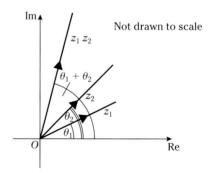

$$\dfrac{z_1}{z_2} = \dfrac{r_1}{r_2} e^{i(\theta_1 - \theta_2)}$$

On an Argand diagram, you can see that when $z_1$ is divided by $z_2$, $z_1$ is enlarged by a scale factor $\dfrac{1}{|z_2|}$ and rotated by an angle $\theta_2$.

The word enlargement is used to describe both getting larger and getting smaller.

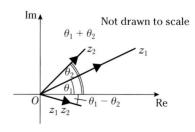

## Examples 15e

1  Given that $z_1 = 3 + 4i$ and $z_2 = 5 - 12i$, find the modulus of

 (a)  $z_1 z_2$

 (b)  $\dfrac{z_1}{z_2}$

 (a)  $|(3 + 4i)(5 - 12i)| = |(3 + 4i)||(5 - 12i)|$

$$= 5 \times 13 = 65$$

 (b)  $\left|\dfrac{3 + 4i}{5 - 12i}\right| = \dfrac{5}{13}$

2  The complex number $\dfrac{1}{\sqrt{2}} + \dfrac{1}{\sqrt{2}}i$ is denoted by $z$.

 (a)  Represent $z$, $iz$ and $\dfrac{1}{iz}$ on an Argand diagram.

 (b)  The complex numbers $z$, $iz$ and $\dfrac{1}{iz}$ are represented by the points $A$, $B$ and $C$ respectively.

 Express the complex number represented by $\overrightarrow{CB}$ in the form $a + ib$.

 (a)  The modulus of $z = \dfrac{1}{\sqrt{2}} + \dfrac{1}{\sqrt{2}}i$ is 1 and the argument is $\dfrac{\pi}{4}$.

 Therefore     $z = e^{i\frac{\pi}{4}}$,     so     $iz = ie^{i\frac{3\pi}{4}}$

 $iz$ is represented by a rotation of $z$ by $\dfrac{\pi}{2}$ so arg $(z) = \dfrac{\pi}{4} + \dfrac{\pi}{2}$

$$\dfrac{1}{iz} = e^{i\left(0 - \frac{3\pi}{4}\right)} = e^{i\left(-\frac{3\pi}{4}\right)}$$

 1 can be expressed as $1e^{0i}$

 (b)  $|z| = |iz| = \left|\dfrac{1}{iz}\right| = 1$

 $z$ can be expressed as $\cos\dfrac{\pi}{4} + i\sin\dfrac{\pi}{4}$,

 $iz$ can be expressed as $\cos\dfrac{3\pi}{4} + i\sin\dfrac{3\pi}{4}$

 $\dfrac{1}{iz}$ can be expressed as $\cos\left(-\dfrac{3\pi}{4}\right) + i\sin\left(-\dfrac{3\pi}{4}\right)$

 $\overrightarrow{CB}$ represents  $iz - \dfrac{1}{iz} = \left(\cos\dfrac{3\pi}{4} + i\sin\dfrac{3\pi}{4}\right) - \left(\cos\left(-\dfrac{3\pi}{4}\right) + i\sin\left(-\dfrac{3\pi}{4}\right)\right).$

$$= \left(-\dfrac{1}{\sqrt{2}} + \dfrac{1}{\sqrt{2}}i\right) - \left(-\dfrac{1}{\sqrt{2}} - \dfrac{1}{\sqrt{2}}i\right)$$

$$= \dfrac{2}{\sqrt{2}}i = i\sqrt{2}$$

 You can get this result direct from the diagram: triangle $OBC$ is a right-angled isosceles triangle, $OB = OC = 1$, so $BC = \sqrt{2}$. Also, $CB$ is parallel to the imaginary axis, so $\overrightarrow{CB}$ represents $i\sqrt{2}$.

## Exercise 15e

1  Find the modulus and argument of:

(a)  $2(1 + i)$

(b)  $(3 - i\sqrt{3})(1 - i)$

(c)  $\dfrac{-2 - i\sqrt{3}}{i\sqrt{3} - 2}$

2  Given that $z = re^{i\theta}$ show that $z^2 = r^2 e^{i2\theta}$

Hence find the square roots of $2\sqrt{3} - 2i$.

Let $z^2 = 2\sqrt{3} - 2i$

3  Given that $z = 3\left(\cos\dfrac{\pi}{4} + i\sin\dfrac{\pi}{4}\right)$,

find the modulus and argument of $z^3$.

4  Convert $z_1 = \dfrac{1}{2} + \dfrac{\sqrt{3}}{2}i$ and $z_2 = 3\dfrac{\sqrt{3}}{2} + \dfrac{3}{2}i$

into polar form.

Hence or otherwise illustrate $z_1 z_2$ and $\dfrac{z_1}{z_2}$

on an Argand diagram.

## LOCI

When $z = x + iy$, the point $P$ representing $z$ can be anywhere on the Argand diagram. When a condition is put on $z$, the positions of $P$ are restricted.

For example, when $|z| = 4$ the line $OP$ is a fixed length of 4 units.

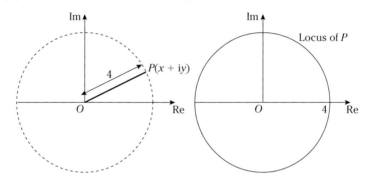

Therefore $P$ is restricted to any point on the circumference of a circle with centre $O$ and radius 4, i.e. $|z| = 4$ defines the circle, centre $O$ and radius 4.

**Any equation of the form $|z| = r$ gives the equation of a circle, centre $O$ and radius $r$.**

When $|z - z_1| = 4$ and $z_1$ is represented by the fixed point $A(x_1 + i y_1)$ and $z$ is represented by $P(x + iy)$ on the Argand diagram, then $z - z_1$ is represented by the line joining $A$ and $P$. So $|z - z_1| = 4$ means that the length of $AP$ is always 4 units.

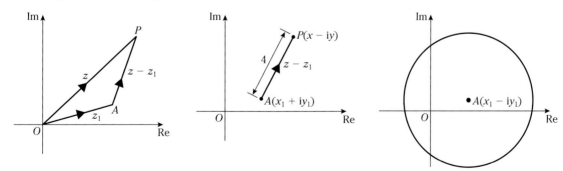

Therefore $P$ is on a circle with radius 4 and centre $A$.

**Any equation of the form $|z - z_1| = a$, where $a$ is a constant, represents a circle, centre $z_1$ and radius $a$.**

## Examples 15f

**1** Sketch on an Argand diagram the locus of points for which $|z - 3| = 2$

Comparing $|z - 3| = 2$ with $|z - z_1| = a$, shows that $z_1 = 3 + 0i$, so the locus is a circle, centre 3 and radius 2.

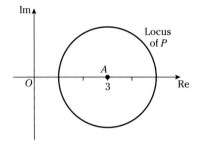

**2** You are given $z_1 = 3 + i$, $z_2 = -3 - i$ and $z = x + iy$

   (a)   Find the locus of the points $P(x + iy)$ on the Argand diagram represented by $|z - z_1| = |z - z_2|$

   (b)   Shade the area on the Argand diagram where $|z - z_1| < |z - z_2|$

   (a)   Taking $A$ as the point $(3 + i)$ and $B$ as the point $(-3 - i)$, $z - z_1 = \overrightarrow{AP}$ and $z - z_2 = \overrightarrow{BP}$

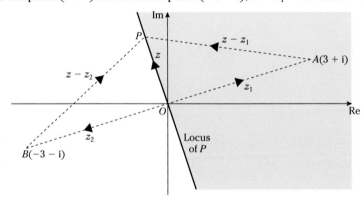

$$\therefore \quad |z - z_1| = |z - z_2| \Rightarrow AP = BP$$

$\therefore$    The locus of $P$ is the perpendicular bisector of $AB$.

   (b)   $|z - z_1| < |z - z_2| \Rightarrow AP < BP$ and this is true for all points to the right of the perpendicular bisector of $AB$.

**3** You are given that $w = r \cos \theta - 3 + ir \sin \theta$

On a sketch of the Argand diagram, shade the area represented by $|w + 3i| < 2$

When $z = r(\cos \theta + i \sin \theta)$, $w = z - 3$

so $w + 3i = z - (3 - 3i)$

$\therefore$    $|w + 3i| < 2$ is equivalent to $|z - (3 - 3i)| < 2$

$|z - (3 - 3i)| = 2$ is represented by a circle, centre $(3 - 3i)$ and radius 2.

$\therefore$    $|z - (3 - 3i)| < 2$ is represented by all the points inside the circle.

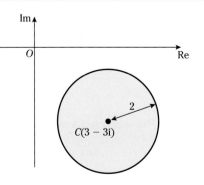

4 Shade the region of the Argand diagram where $0 \leqslant \arg (z - 3 + 2i) \leqslant \dfrac{\pi}{2}$

In the diagram, $w = z - (3 - 2i)$

The diagram shows that $\arg w = 0$ when $z = \overrightarrow{OP}$ where $P$ is any point on the line through $A$ parallel to the real axis.

Also $\arg w = \dfrac{\pi}{2}$ when $P$ is any point on the line through $A$ parallel to the imaginary axis.

For any position of $P$ between these two lines,

$$0 \leqslant \arg (z - 3 + 2i) \leqslant \dfrac{\pi}{2}$$

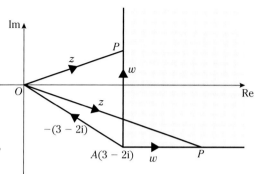

5 Sketch on an Argand diagram the locus of points where $|z - 2| = |z - (1 - 2i)|$

In the diagram $\quad \overrightarrow{AP} = z - 2$

and $\quad\quad\quad\quad \overrightarrow{BP} = z - (1 - 2i)$

$AP = BP$ when $P$ is on the perpendicular bisector of $AB$.

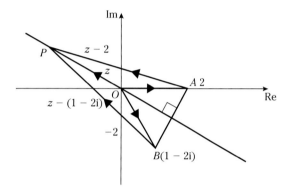

6 (a) Find the complex number represented by the point, $A$, of intersection of the loci $|z| = 2$ and $\arg (z) = \dfrac{\pi}{3}$

(b) Find points $B$ and $C$ on $|z| = 2$ such that triangle $ABC$ is equilateral.

(a) From the diagram, $z$ has modulus 2 and argument $\dfrac{\pi}{3}$

$\quad \therefore \quad\quad z = 2 \cos \dfrac{\pi}{3} + 2i \sin \dfrac{\pi}{3}$

$\quad\quad\quad\quad = 1 + i\sqrt{3}$

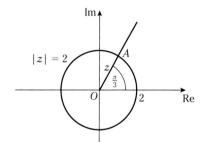

(b) $A$, $B$ and $C$ are on the circumference of the circle. The centre $O$ of the circle is the same distance from all three vertices.
For triangle $ABC$ to be equilateral, angles $AOB$, $BOC$, $COA$ are all equal to $\dfrac{2\pi}{3}$.

Therefore $C$ represents $-2$ and $B$ represents $1 - i\sqrt{3}$

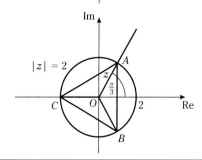

## Exercise 15f

Sketch the locus on an Argand diagram of points for which

1  $|z| = 1$

2  $|z - 1| = 3$

3  $|z - 2i| = 3$

4  $|z + 2| = 2$

5  $|z - 1 + i| = 4$

6  $\arg z = \dfrac{\pi}{3}$

7  $|z - 2 - 3i| = |z + 4 - 5i|$

8  $|z| = |z + 4i|$

9  Shade on an Argand diagram the areas represented by the following inequalities.

   (a)  $|z - 1| < 4$       (b)  $|z + 3i| > 2$

   (c)  $|z + 1 - i| < 1$     (d)  $\dfrac{\pi}{3} < \arg z < \dfrac{2\pi}{3}$

10  Shade on an Argand diagram the region occupied by the set of points $P(x, y)$ for which $|z| < 5$ *and* $-\dfrac{\pi}{6} < \arg z < \dfrac{\pi}{6}$, where $z = x + iy$

11  Show, on an Argand diagram, the set of points for which

   (a)  $|z| = 4$    and    $\arg z = \dfrac{\pi}{4}$

   (b)  $|z + 2 + i| = 5$    and    $\text{Re}(z - 1) = 0$

12  Find the complex number represented by the points of intersection of the loci on an Argand diagram defined by
$$\arg z = -\dfrac{\pi}{4} \quad \text{and} \quad |z| = 2$$

13  Indicate on an Argand diagram the set of points $P(x, y)$ for which

   (a)  $0 \leqslant \arg (z + 1) \leqslant \dfrac{\pi}{3}$   and   $|z + i| = 3$

   (b)  $|z + 3 - 2i| < 4$   and   $\arg (z + 1) = \dfrac{5\pi}{6}$

   (c)  $|z| > 1, |z| < 4$   and   $\arg z = -\dfrac{3\pi}{4}$

## Mixed exercise 15

1  If $z_1 = \dfrac{2 - i}{2 + i}$, $z_2 = \dfrac{2i - 1}{1 - i}$, express $z_1$ and $z_2$ in the form $a + ib$.

   Sketch an Argand diagram showing points $P$ and $Q$ representing the complex numbers $5z_1 + 2z_2$ and $5z_1 - 2z_2$ respectively.

2  If $(1 + 3i)z_1 = 5(1 + i)$, express $z_1$ and $z_1{}^2$ in the form $x + iy$, where $x$ and $y$ are real.

   Sketch in an Argand diagram the circle $|z - z_1| = |z_1|$, giving the coordinates of its centre.

3  (a)  If $z = 4 - 3i$ express $z + \dfrac{1}{z}$ in the form $a + ib$.

   (b)  Find the two square roots of $4i$ in the form $a + ib$.

   (c)  If $z_1 = 5 - 5i$ and $z_2 = -1 + 7i$ prove that:
   $$|z_1 + z_2| < |z_1 - z_2| < |z_1| + |z_2|$$

4  Express the complex number $\dfrac{5 + 12i}{3 + 4i}$ in the form $a + ib$ and in the form $r(\cos \theta + i \sin \theta)$, giving the values of $a, b, r, \cos \theta, \sin \theta$.

5  The complex numbers $z_1 = \dfrac{a}{1 + i}$, $z_2 = \dfrac{b}{1 + 2i}$ where $a$ and $b$ are real, are such that $z_1 + z_2 = 1$. Find $a$ and $b$.

   With these values of $a$ and $b$, find the distance between the points that represent $z_1$ and $z_2$ in the Argand diagram.

6  Find the modulus and argument of $z_1 = \sqrt{3} + i$. If $z_2 = \sqrt{3} - i$ express $q = \dfrac{z_1}{z_2}$ in the form $a + ib$ where $a$ and $b$ are real.

7  (a)  The complex number $z$ and its conjugate $z^*$ satisfy the equation
   $$zz^* + 2iz = 12 + 6i$$
   Find the values of $z$.

   (b)  Mark on an Argand diagram the points representing $4 + 3i$, $4 - 3i$ and $\dfrac{4 + 3i}{4 - 3i}$

8 (a) If $z = 3 + 4i$, express $z + \dfrac{25}{z}$ in its simplest form.

(b) If $z = x + iy$, find the real part and the imaginary part of $z + \dfrac{1}{z}$.

Find the locus of points in the Argand diagram for which the imaginary part of $z + \dfrac{1}{z}$ is zero.

9 (a) Find the square roots of $(5 + 12i)$.

(b) Find the modulus and argument of each of the numbers

   (i) $(1 - i)$

   (ii) $(4 + 3i)$

   (iii) $(1 - i)(4 + 3i)$.

If these numbers are represented in an Argand diagram by the points $A$, $B$, $C$, calculate the area of the triangle $ABC$.

(c) Find the ratio of the greatest value of $|z + 1|$ to its least value when $|z - i| = 1$

10 (a) Find the modulus and one value for the argument of $\dfrac{(i + 1)^2}{(i - 1)^4}$

(b) Find the two square roots of $5 - 12i$ in the form $a + ib$ where $a$ and $b$ are real. Show the points $P$ and $Q$ representing the square roots in an Argand diagram. Find the complex numbers represented by points $R_1$, $R_2$ such that the triangles $PQR_1$, $PQR_2$ are equilateral.

11 Prove that the modulus of $2 + \cos\theta + i\sin\theta$ is $(5 + 4\cos\theta)^{\frac{1}{2}}$

Hence show that the modulus of

$$\dfrac{2 + \cos\theta + i\sin\theta}{2 + \cos\theta - i\sin\theta}$$

is unity.

# Summary 3

## ALGEBRA

### Partial fractions

A proper fraction with a denominator that factorises can be expressed in partial fractions as follows:

$$\frac{f(x)}{(x-a)(x-b)} = \frac{A}{(x-a)} + \frac{B}{(x-b)} \quad \text{and} \quad \frac{f(x)}{(x-a)(x-b)(x-c)} = \frac{A}{x-a} + \frac{B}{x-b} + \frac{C}{x-c}$$

$$\frac{f(x)}{(x-a)(x-b)^2} = \frac{A}{(x-a)} + \frac{B}{(x-b)} + \frac{C}{(x-b)^2}$$

$$\frac{f(x)}{(x-a)(x^2+b)} = \frac{A}{(x-a)} + \frac{Bx+C}{(x^2+b)}$$

### The binomial theorem

If $n$ is *any* number then $(1+x)^n$ can be expanded as the *infinite* series given by

$$(1+x)^n = 1 + nx + \binom{n}{2}x^2 + \binom{n}{3}x^3 + \dots$$

provided that $-1 < x < 1$, where $\binom{n}{r} = \dfrac{n(n-1) \dots (n-r+1)}{r!}$

## INTEGRATION

**Integrating products can be done by**

(a) recognition: in particular

$$\int f'(x)e^{f(x)}\,dx = e^{f(x)} + K$$

(b) substitution

(c) parts: $\displaystyle\int v\,\frac{du}{dx}\,dx = uv - \int u\,\frac{dv}{dx}\,dx$

Integration by parts can be used also to integrate $\ln x$.

**Integrating fractions can be done by**

(a) recognition: in particular

$$\int \frac{f'(x)}{f(x)}\,dx = \ln|f(x)| + K$$

(b) substitution

(c) using partial fractions.

## DIFFERENTIAL EQUATIONS

A first order linear differential equation is a relationship between $x$, $y$ and $\dfrac{dy}{dx}$. It can be solved by collecting all the $x$ terms, along with $dx$, on one side, with all $y$ terms and $dy$ on the other side. Then each side is integrated with respect to its own variable. A constant of integration called an arbitrary constant is introduced on one side only to give a general solution which is a family of lines or curves.

# VECTORS

## Equations for a line

For a line in the direction of the vector $\mathbf{d} = a\mathbf{i} + b\mathbf{j} + c\mathbf{k}$ and passing through a point with position vector $\mathbf{a} = x_1\mathbf{i} + y_1\mathbf{j} + z_1\mathbf{k}$, a vector equation in standard form is $\mathbf{r} = \mathbf{a} + t\mathbf{d}$

Two lines with equations $\mathbf{r}_1 = \mathbf{a}_1 + t\mathbf{d}_1$ and $\mathbf{r}_2 = \mathbf{a}_2 + s\mathbf{d}_2$

   are parallel if $\mathbf{d}_1$ is a multiple of $\mathbf{d}_2$

   intersect if there are values of $t$ and $s$ for which $\mathbf{r}_1 = \mathbf{r}_2$

   are skew in all other cases.

## Vector equations of a plane

A plane perpendicular to a vector $\mathbf{n}$ has a vector equation in standard form of $(\mathbf{r} - \mathbf{a}).\mathbf{n} = 0$

where $\mathbf{a}$ is a position vector of a point on the plane and $\mathbf{n}$ is a vector perpendicular to the plane.

A Cartesian equation of a plane is $ax + by + cz = d$ where $(a, b, c)$ is a point on the plane.

The perpendicular distance of a plane from $O$ is $\dfrac{\mathbf{a}.\mathbf{n}}{|\mathbf{n}|}$ or $\dfrac{d}{\sqrt{a^2 + b^2 + c^2}}$

# COMPLEX NUMBERS

$x + iy = a + bi \Leftrightarrow x = a, y = b$

$|a + bi| = \sqrt{a^2 + b^2}$, $\arg(a + bi) = \theta$ where $\tan\theta = \dfrac{b}{a}$ and $-\pi < \theta < \pi$

$z = x + iy$ then $z^* = x - iy$

$z = r(\cos\theta + i\sin\theta) = re^{i\theta}, r = |z|, \theta = \arg z$

$|z_1 z_2| = |z_1||z_2|$ and $\arg z_1 z_2 = \arg z_1 + \arg z_2$

$\left|\dfrac{z_1}{z_2}\right| = \dfrac{|z_1|}{|z_2|}$ and $\arg\dfrac{z_1}{z_2} = \arg z_1 - \arg z_2$

## Summary exercise 3

1  The complex number $2i$ is denoted by $u$. The complex number with modulus 1 and argument $\dfrac{2\pi}{3}$ is denoted by $w$.

   (i) Find in the form $x + iy$, where $x$ and $y$ are real, the complex numbers $w$, $uw$ and $\dfrac{u}{w}$. [4]

   (ii) Sketch an Argand diagram showing the points $U$, $A$ and $B$ representing the complex numbers $u$, $uw$ and $\dfrac{u}{w}$ respectively. [2]

   (iii) Prove that triangle $UAB$ is equilateral. [2]
   Cambridge, Paper 3 Q5 J03

2  Let $f(x) = \dfrac{9x^2 + 4}{(2x + 1)(x - 2)^2}$

   (i) Express $f(x)$ in partial fractions. [5]

   (ii) Show that, when $x$ is sufficiently small for $x^3$ and higher powers to be neglected,
   $$f(x) = 1 - x + 5x^2 \qquad [4]$$
   Cambridge, Paper 3 Q6 J03

3  In a chemical reaction a compound $X$ is formed from a compound $Y$. The masses in grams of $X$ and $Y$ present at time $t$ seconds after the start of the reaction are $x$ and $y$ respectively. The sum of the two masses is equal to 100 grams throughout the reaction. At any time, the rate of formation of $X$ is proportional to the mass of $Y$ at that time. When $t = 0$, $x = 5$ and $\dfrac{dx}{dt} = 1.9$

   (i) Show that $x$ satisfies the differential equation
   $$\frac{dx}{dt} = 0.02(100 - x) \qquad [2]$$

(ii) Solve this differential equation, obtaining an expression for $x$ in terms of $t$.    [6]

(iii) State what happens to the value of $x$ as $t$ becomes very large.    [1]

Cambridge, Paper 3 Q7 J03

4  Two planes have equations $x + 2y - 2z = 2$ and $2x - 3y + 6z = 3$. The planes intersect in the straight line $l$.

(i) Calculate the acute angle between the two planes.    [4]

(ii) Find a vector equation for the line $l$.    [6]

Cambridge, Paper 3 Q9 J03

5  Given that $y = 1$ when $x = 0$, solve the differential equation

$$\frac{dy}{dx} = \frac{y^3 + 1}{y^2}$$

obtaining an expression for $y$ in terms of $x$.    [6]

Cambridge, Paper 3 Q6 J04

6  (i) Find the roots of the equation $z^2 - z + 1 = 0$, giving your answers in the form $x + iy$, where $x$ and $y$ are real.    [2]

(ii) Obtain the modulus and argument of each root.    [3]

(iii) Show that each root also satisfies the equation $z^3 = -1$    [2]

Cambridge, Paper 3 Q8 J04

7  Let $f(x) = \dfrac{x^2 + 7x - 6}{(x - 1)(x - 2)(x + 1)}$

(i) Express $f(x)$ in partial fractions.    [4]

(ii) Show that, when $x$ is sufficiently small for $x^4$ and higher powers to be neglected,

$$f(x) = -3 + 2x - \tfrac{3}{2}x^2 + \tfrac{11}{4}x^3$$    [5]

Cambridge, Paper 3 Q9 J04

8

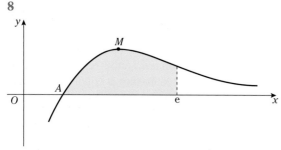

The diagram shows the curve $y = \dfrac{\ln x}{x^2}$ and its maximum point $M$. The curve cuts the $x$-axis at $A$.

(i) Write down the $x$-coordinate of $A$.    [1]

(ii) Find the exact coordinates of $M$.    [5]

(iii) Use integration by parts to find the exact area of the shaded region enclosed by the curve, the $x$-axis and the line $x = e$    [5]

Cambridge, Paper 3 Q10 J04

9  Expand $(2 + 3x)^{-2}$ in ascending powers of $x$, up to and including the term in $x^2$, simplifying the coefficients.    [4]

Cambridge, Paper 3 Q1 J07

10  The equation of a curve is $y = x \sin 2x$, where $x$ is in radians. Find the equation of the tangent to the curve at the point where $x = \tfrac{1}{4}\pi$    [4]

Cambridge, Paper 3 Q3 J07

11  Using the substitution $u = 3^x$, or otherwise, solve, correct to 3 significant figures, the equation

$$3^x = 2 + 3^{-x}$$    [6]

Cambridge, Paper 3 Q4 J07

12  (i) Express $\cos \theta + \sqrt{3} \sin \theta$ in the form $R \cos (\theta - \alpha)$, where $R > 0$ and $0 < \alpha < \tfrac{1}{2}\pi$, giving the exact values of $R$ and $\alpha$.    [3]

(ii) Hence show that

$$\int_0^{\frac{1}{2}\pi} \frac{1}{(\cos \theta + (\sqrt{3}) \sin \theta)^2} \, d\theta = \frac{1}{\sqrt{3}}$$    [4]

Cambridge, Paper 3 Q5 J07

13  A model for the height, $h$ metres, of a certain type of tree at time $t$ years after being planted assumes that, while the tree is growing, the rate of increase in height is proportional to $(9 - h)^{\frac{1}{3}}$. It is given that, when $t = 0$, $h = 1$ and $\dfrac{dh}{dt} = 0.2$

(i) Show that $h$ and $t$ satisfy the differential equation

$$\frac{dh}{dt} = 0.1(9 - h)^{\frac{1}{3}}$$    [2]

(ii) Solve this differential equation, and obtain an expression for $h$ in terms of $t$.    [7]

(iii) Find the maximum height of the tree and the time taken to reach this height after planting.    [2]

(iv) Calculate the time taken to reach half the maximum height.    [1]

Cambridge, Paper 3 Q10 J07

**14** The points $A$ and $B$ have position vectors, relative to the origin $O$, given by

$$\overrightarrow{OA} = \mathbf{i} + 2\mathbf{j} + 3\mathbf{k} \quad \text{and} \quad \overrightarrow{OB} = 2\mathbf{i} + \mathbf{j} + 3\mathbf{k}$$

The line $l$ has vector equation

$$\mathbf{r} = (1 - 2t)\mathbf{i} + (5 + t)\mathbf{j} + (2 - t)\mathbf{k}$$

(i) Show that $l$ does not intersect the line passing through $A$ and $B$. [4]

(ii) The point $P$ lies on $l$ and is such that angle $PAB$ is equal to $60°$. Given that the position vector of $P$ is $(1 - 2t)\mathbf{i} + (5 + t)\mathbf{j} + (2 - t)\mathbf{k}$, show that $3t^2 + 7t + 2 = 0$. Hence find the only possible position vector of $P$. [6]
*Cambridge, Paper 3 Q10 J08*

**15**
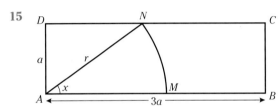

In the diagram, $ABCD$ is a rectangle with $AB = 3a$ and $AD = a$. A circular arc, with centre $A$ and radius $r$, joins points $M$ and $N$ on $AB$ and $CD$ respectively. The angle $MAN$ is $x$ radians. The perimeter of the sector $AMN$ is equal to half the perimeter of the rectangle.

(i) Show that $x$ satisfies the equation

$$\sin x = \tfrac{1}{4}(2 + x)$$ [3]

(ii) This equation has only one root in the interval $0 < x < \tfrac{1}{2}\pi$. Use the iterative formula

$$x_{n+1} = \sin^{-1}\left(\frac{2 + x_n}{4}\right)$$

with initial value $x_1 = 0.8$ to determine the root correct to 2 decimal places. Give the result of each iteration to 4 decimal places. [3]
*Cambridge, Paper 3 Q3 J08*

**16** (i) Show that the equation $\tan(30° + \theta) = 2\tan(60° - \theta)$ can be written in the form

$$\tan^2\theta + (6\sqrt{3})\tan\theta - 5 = 0$$ [4]

(ii) Hence, or otherwise, solve the equation

$$\tan(30° + \theta) = 2\tan(60° - \theta)$$

for $0° \leq \theta \leq 180°$ [3]
*Cambridge, Paper 3 Q4 J08*

**17** When $(1 + 2x)(1 + ax)^{\frac{2}{3}}$, where $a$ is a constant, is expanded in ascending powers of $x$, the coefficient of the term in $x$ is zero.

(i) Find the value of $a$. [3]

(ii) When $a$ has this value, find the term in $x^3$ in the expansion of $(1 + 2x)(1 + ax)^{\frac{2}{3}}$, simplifying the coefficient. [4]
*Cambridge, Paper 3 Q5 J09*

**18**
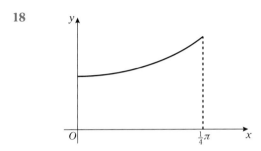

The diagram shows the curve
$y = \sqrt{(1 + 2\tan^2 x)}$ for $0 \leq x \leq \tfrac{1}{4}\pi$

(i) Use the trapezium rule with three intervals to estimate the value of

$$\int_0^{\frac{1}{4}\pi} \sqrt{(1 + 2\tan^2 x)} \, dx,$$

giving your answer correct to 2 decimal places. [3]

(ii) The estimate found in part (i) is denoted by $E$. Explain, without further calculation, whether another estimate found using the trapezium rule with six intervals would be greater than $E$ or less than $E$. [1]
*Cambridge, Paper 3 Q2 J09*

**19** (i) Solve the equation $z^2 + (2\sqrt{3})iz - 4 = 0$, giving your answers in the form $x + iy$, where $x$ and $y$ are real. [3]

(ii) Sketch an Argand diagram showing the points representing the roots. [1]

(iii) Find the modulus and argument of each root. [3]

(iv) Show that the origin and the points representing the roots are the vertices of an equilateral triangle. [1]
*Cambridge, Paper 3 Q7 J09*

**20** (i) Express $\dfrac{100}{x^2(10 - x)}$ in partial fractions. [4]

(ii) Given that $x = 1$ when $t = 0$, solve the differential equation

$$\frac{dx}{dt} = \frac{1}{100}x^2(10 - x)$$

obtaining an expression for $t$ in terms of $x$.    [6]

Cambridge, Paper 3 Q8 J09

**21** The line $l$ has equation
$\mathbf{r} = 4\mathbf{i} + 2\mathbf{j} - \mathbf{k} + t(2\mathbf{i} - \mathbf{j} - 2\mathbf{k})$. It is given that $l$ lies in the plane with equation $2x + by + cz = 1$, where $b$ and $c$ are constants.

(i) Find the values of $b$ and $c$.    [6]

(ii) The point $P$ has position vector $2\mathbf{j} + 4\mathbf{k}$. Show that the perpendicular distance from $P$ to $l$ is $\sqrt{5}$.    [5]

Cambridge, Paper 3 Q9 J09

**22** The equation of a curve is

$$x \ln y = 2x + 1$$

(i) Show that $\dfrac{dy}{dx} = -\dfrac{y}{x^2}$    [4]

(ii) Find the equation of the tangent to the curve at the point where $y = 1$, giving your answer in the form $ax + by + c = 0$    [4]

Cambridge, Paper 32 Q6 J10

**23** The variable complex number $z$ is given by

$$z = 1 + \cos 2\theta + i \sin 2\theta$$

where $\theta$ takes all values in the interval $-\frac{1}{2}\pi < \theta < \frac{1}{2}\pi$

(i) Show that the modulus of $z$ is $2 \cos \theta$ and the argument of $z$ is $\theta$.    [6]

(ii) Prove that the real part of $\dfrac{1}{z}$ is constant.    [3]

Cambridge, Paper 32 Q8 J10

**24** The plane $p$ has equation $3x + 2y + 4z = 13$. A second plane $q$ is perpendicular to $p$ and has equation $ax + y + z = 4$, where $a$ is a constant.

(i) Find the value of $a$.    [3]

(ii) The line with equation
$\mathbf{r} = \mathbf{j} - \mathbf{k} + \lambda(\mathbf{i} + 2\mathbf{j} + 2\mathbf{k})$ meets the plane $p$ at the point $A$ and the plane $q$ at the point $B$. Find the length of $AB$.    [6]

Cambridge, Paper 32 Q9 J10

**25** (i) Find the values of the constants $A$, $B$, $C$ and $D$ such that

$$\frac{2x^3 - 1}{x^2(2x - 1)} \equiv A + \frac{B}{x} + \frac{C}{x^2} + \frac{D}{2x - 1}$$

[5]

(ii) Hence show that

$$\int_1^2 \frac{2x^3 - 1}{x^2(2x - 1)}\, dx = \frac{3}{2} + \frac{1}{2}\ln\left(\frac{16}{27}\right)$$    [5]

Cambridge, Paper 32 Q10 J10

**26** In a model of the expansion of a sphere of radius $r$ cm it is assumed that, at time $t$ seconds after the start, the rate of increase of the surface area of the sphere is proportional to its volume. When $t = 0$, $r = 5$ and $\dfrac{dr}{dt} = 2$

(i) Show that $r$ satisfies the differential equation

$$\frac{dr}{dt} = 0.08r^2$$    [4]

[The surface area $A$ and volume $V$ of a sphere of radius $r$ are given by the formulae $A = 4\pi r^2$, $V = \frac{4}{3}\pi r^3$]

(ii) Solve this differential equation, obtaining an expression for $r$ in terms of $t$.    [5]

(iii) Deduce from your answer to part (ii) the set of values that $t$ can take, according to this model.    [1]

Cambridge, Paper 31 Q10 N09

**27** A curve has equation $y = e^{-3x} \tan x$. Find the $x$-coordinates of the stationary points on the curve in the interval $-\frac{1}{2}\pi < x < \frac{1}{2}\pi$. Give your answers correct to 3 decimal places.    [6]

Cambridge, Paper 31 Q4 N09

**28** The sequence of values given by the iterative formula

$$x_{n+1} = \frac{3x_n}{4} + \frac{15}{x_n^3}$$

with initial value $x_1 = 3$, converges to $\alpha$.

(i) Use this iterative formula to find $\alpha$ correct to 2 decimal places, giving the result of each iteration to 4 decimal places.    [3]

(ii) State an equation satisfied by $\alpha$ and hence find the exact value of $\alpha$.    [2]

Cambridge, Paper 31 Q3 N09

29  (i) Express $\dfrac{5x + 3}{(x + 1)^2(3x + 2)}$ in partial fractions.    [5]

   (ii) Hence obtain the expansion of $\dfrac{5x + 3}{(x + 1)^2(3x + 2)}$ in ascending powers of $x$, up to and including the term in $x^2$, simplifying the coefficients.    [5]

   Cambridge, Paper 31 Q8 N09

30  The constant $a$ is such that $\displaystyle\int_0^a xe^{\frac{1}{2}x}\, dx = 6$

   (i) Show that $a$ satisfies the equation

   $$x = 2 + e^{-\frac{1}{2}x}$$    [5]

   (ii) By sketching a suitable pair of graphs, show that this equation has only one root.    [2]

   (iii) Verify by calculation that this root lies between 2 and 2.5.    [2]

   (iv) Use an iterative formula based on the equation in part (i) to calculate the value of $a$ correct to 2 decimal places. Give the result of each iteration to 4 decimal places.    [3]

   Cambridge, Paper 3 Q9 N08

31  Two planes have equations $2x - y - 3z = 7$ and $x + 2y + 2z = 0$

   (i) Find the acute angle between the planes.    [4]

   (ii) Find a vector equation for their line of intersection.    [6]

   Cambridge, Paper 3 Q7 N08

32  Expand $(1 + x)\sqrt{(1 - 2x)}$ in ascending powers of $x$, up to and including the term in $x^2$, simplifying the coefficients.    [4]

   Cambridge, Paper 3 Q2 N08

33  The polynomial $4x^3 - 4x^2 + 3x + a$, where $a$ is a constant, is denoted by $p(x)$. It is given that $p(x)$ is divisible by $2x^2 - 3x + 3$.

   (i) Find the value of $a$.    [3]

   (ii) When $a$ has this value, solve the inequality $p(x) < 0$, justifying your answer.    [3]

   Cambridge, Paper 3 Q5 N08

# List of formulae

*Algebra*

For the quadratic equation $ax^2 + bx + c = 0$:

$$x = \frac{-b \pm \sqrt{b^2 - 4ac}}{2a}$$

For an arithmetic series:

$$u_n = a + (n-1)d, \qquad S_n = \tfrac{1}{2}n(a+l) = \tfrac{1}{2}n\{2a + (n-1)d\}$$

For a geometric series:

$$u_n = ar^{n-1}, \qquad S_n = \frac{a(1-r^n)}{1-r} \ (r \neq 1), \qquad S_\infty = \frac{a}{1-r} \qquad (|r| < 1)$$

Binomial expansion:

$$(a+b)^n = a^n + \binom{n}{1}a^{n-1}b + \binom{n}{2}a^{n-2}b^2 + \binom{n}{3}a^{n-3}b^3 + \ldots + b^n, \text{ where } n \text{ is a positive integer}$$

$$\text{and } \binom{n}{r} = \frac{n!}{r!(n-r)!}$$

$$(1+x)^n = 1 + nx + \frac{n(n-1)}{2!}x^2 + \frac{n(n-1)(n-2)}{3!}x^3 \ldots, \text{ where } n \text{ is rational and } |x| < 1$$

*Trigonometry*

Arc length of circle $= r\theta$ ($\theta$ in radians)

Area of sector of circle $= \tfrac{1}{2}r^2\theta$ ($\theta$ in radians)

$$\tan\theta \equiv \frac{\sin\theta}{\cos\theta}$$

$$\cos^2\theta + \sin^2\theta \equiv 1, \qquad 1 + \tan^2\theta \equiv \sec^2\theta, \qquad \cot^2\theta + 1 \equiv \operatorname{cosec}^2\theta$$

$$\sin(A \pm B) \equiv \sin A \cos B \pm \cos A \sin B$$

$$\cos(A \pm B) \equiv \cos A \cos B \mp \sin A \sin B$$

$$\tan(A \pm B) \equiv \frac{\tan A \pm \tan B}{1 \mp \tan A \tan B}$$

$$\sin 2A \equiv 2 \sin A \cos A$$

$$\cos 2A \equiv \cos^2 A - \sin^2 A \equiv 2\cos^2 A - 1 \equiv 1 - 2\sin^2 A$$

$$\tan 2A \equiv \frac{2\tan A}{1 - \tan^2 A}$$

Principal values:

$$-\tfrac{1}{2}\pi \leqslant \sin^{-1}x \leqslant \tfrac{1}{2}\pi$$

$$0 \leqslant \cos^{-1}x \leqslant \pi$$

$$-\tfrac{1}{2}\pi \leqslant \tan^{-1}x < \tfrac{1}{2}\pi$$

*Differentiation*

| $f(x)$ | $f'(x)$ |
|---|---|
| $x^n$ | $nx^{n-1}$ |
| $\ln x$ | $\dfrac{1}{x}$ |
| $e^x$ | $e^x$ |
| $\sin x$ | $\cos x$ |
| $\cos x$ | $-\sin x$ |
| $\tan x$ | $\sec^2 x$ |
| $uv$ | $u\dfrac{dv}{dx} + v\dfrac{du}{dx}$ |
| $\dfrac{u}{v}$ | $\dfrac{v\dfrac{du}{dx} - u\dfrac{dv}{dx}}{v^2}$ |

If $x = f(t)$ and $y = g(t)$ then $\dfrac{dy}{dx} = \dfrac{dy}{dt} \div \dfrac{dx}{dt}$

*Integration*

| $f(x)$ | $\displaystyle\int f(x)\,dx$ |
|---|---|
| $x^n$ | $\dfrac{x^{n+1}}{n+1} + c \ (n \neq -1)$ |
| $\dfrac{1}{x}$ | $\ln|x| + c$ |
| $e^x$ | $e^x + c$ |
| $\sin x$ | $-\cos x + c$ |
| $\cos x$ | $\sin x + c$ |
| $\sec^2 x$ | $\tan x + c$ |

$$\int u\frac{dv}{dx}\,dx = uv - \int v\frac{du}{dx}\,dx$$

$$\int \frac{f'(x)}{f(x)}\,dx = \ln|f(x)| + c$$

*Vectors*

If $\mathbf{a} = a_1\mathbf{i} + a_2\mathbf{j} + a_3\mathbf{k}$     and     $\mathbf{b} = b_1\mathbf{i} + b_2\mathbf{j} + b_3\mathbf{k}$ then

$$\mathbf{a}.\mathbf{b} = a_1b_1 + a_2b_2 + a_3b_3 = |\mathbf{a}||\mathbf{b}|\cos\theta$$

*Numerical integration*

Trapezium rule:

$$\int_a^b f(x)\,dx \approx \tfrac{1}{2}h\{y_0 + 2(y_1 + y_2 + \ldots + y_{n-1}) + y_n\}, \text{ where } h = \frac{b-a}{n}$$

# P2 Sample examination papers

Time allowed 1 hour 15 minutes

*Answer **all** the questions. Give non-exact numerical answers correct to 3 significant figures, or 1 decimal place in the case of angles in degrees, unless a different level of accuracy is specified in the question. The use of an electronic calculator is expected, where appropriate. You are reminded of the need for clear presentation in your answers.*

*The number of marks is given in brackets [ ] at the end of each question or part question.*

*The total number of marks for this paper is 50.*

*Note: The number of marks for each question reflects the amount of working required in the answer.*

## Sample paper 1

**Q1** Use logarithms to solve the equation $4^x = 3^{2x+1}$, giving your answer correct to 3 significant figures. [4]

**Q2** Solve the inequality $|2x + 2| < |x - 4|$ [5]

**Q3** Show that $\int_0^1 (e^{2x} - 1)^2 \, dx = 1\frac{3}{4} + \frac{1}{4}e^4 - e^2$ [5]

**Q4** The parametric equations of a curve are

$$x = \cos t, \qquad y = \sin t \qquad \text{where} \qquad 0 \leqslant t < 2\pi$$

Find the equation of the normal at the point where $t = \frac{1}{6}\pi$ [6]

**Q5** (i) Solve the equation $2\sin^2 \theta = \cos \theta + 1$, giving all values in the interval $0 \leqslant \theta \leqslant 2\pi$ [6]

(ii) Prove the identity $\dfrac{1 - \sin^2 \theta}{\csc^2 \theta - 1} = 1 - \cos^2 \theta$ [3]

**Q6** (i) Show that the $x$-coordinate of the intersection of the graphs of $y = \dfrac{8}{x^2}$ and $y = 2x - 1$ satisfies the equation $2x^3 - x^2 - 8 = 0$ [2]

(ii) Verify that the $x$-coordinate lies between 1.6 and 1.8. [2]

(iii) Show that the equation can be rearranged to give $x = \sqrt{\dfrac{8}{2x - 1}}$ [2]

(iv) Use the iterative formula $x_{n+1} = \sqrt{\dfrac{8}{2x_n - 1}}$ with initial value $x_1 = 1.75$ to find the $x$-coordinate correct to 3 significant figures. Give the result of each iteration correct to 5 decimal places. [3]

**Q7** (i) Draw a sketch of the graph of $y = \ln x$ for $0 < x \leqslant 5$ [2]

(ii) Use the trapezium rule with 4 strips to estimate $\int_1^5 \ln x \, dx$, giving your answer correct to 2 decimal places, and stating whether this answer is an underestimate or an overestimate. [5]

(iii) Find $\dfrac{d}{dx}(x \ln x)$ [2]

(iv) Find the exact value of $\int_1^5 \ln x \, dx$ [3]

## Sample paper 2

**Q1** Solve the equation $\log_2(x^2 + 12) - \log_2 x = 3$ [5]

**Q2** (i) The curve $C$ has equation $y = 2 \sec 2x$ for $0 \leqslant x < \frac{1}{4}\pi$. Sketch the graph of $C$, indicating clearly any intersections with the coordinate axes and any asymptotes. [2]

(ii) Find the volume generated when the region bounded by $C$, the coordinate axes and the line $x = \frac{1}{8}\pi$ is rotated through $2\pi$ radians about the $x$-axis. [4]

**Q3** The polynomial $2x^3 + 15x^2 + ax - 8$, where $a$ is a constant, is divisible by $(x - 1)$.
Find the value of $a$, and hence find the solutions of the equation $2x^3 + 15x^2 + ax - 8 = 0$ [7]

**Q4** (i) Express $4 \cos x° + 3 \sin x°$ in the form $R \cos(x - \alpha)°$ where $R > 0$ and $0 < \alpha < 90$ [3]

(ii) Hence solve the equation $4 \cos x° + 3 \sin x° = 2$, giving all solutions between 0 and 360. [4]

**Q5** (i) By sketching the graphs with equations $y = e^{-x}$ and $y = x^2 - 1$ on the same set of axes, show that there is a solution of the equation $e^{-x} = x^2 - 1$ between 1 and 2. [3]

(ii) Show that the equation can be rearranged to give $x = \sqrt{1 + e^{-x}}$ [1]

(iii) Use the iterative formula $x_{n+1} = \sqrt{1 + e^{-x_n}}$ with the initial value $x_1 = 1.1$ to find the solution correct to 3 decimal places. Give the result of each iteration correct to 4 decimal places. [3]

**Q6** (i) Find $\dfrac{\mathrm{d}}{\mathrm{d}x} \sqrt{2 - \cos x}$ [3]

(ii) Hence show that $\displaystyle\int_0^{\frac{1}{3}\pi} \dfrac{\sin x}{\sqrt{2 - \cos x}} \, \mathrm{d}x = \sqrt{6} - 2$ [5]

**Q7** The parametric equations of a curve are

$$x = t e^{-2t}, \quad y = t^2 e^{-2t}$$

Show that the equation of the tangent at the point where $t = -1$ is $4x + 3y + e^2 = 0$ [10]

# P3 Sample examination papers

*Time allowed 1 hour 45 minutes*

*Answer **all** the questions. Give non-exact numerical answers correct to 3 significant figures, or 1 decimal place in the case of angles in degrees, unless a different level of accuracy is specified in the question. The use of an electronic calculator is expected, where appropriate. You are reminded of the need for clear presentation in your answers.*

*The number of marks is given in brackets [ ] at the end of each question or part question.*

*The total number of marks for this paper is 50.*

*Note: The number of marks for each question reflects the amount of working required in the answer.*

## Sample paper 1

**Q1** Solve the inequality $2|x - 3| > |3x + 1|$ [4]

**Q2** Solve the simultaneous equations

$$\log(x + y) = 0$$

$$2 \log x = \log(5 + y)$$ [5]

**Q3** Prove the identity $(\cot \theta + \operatorname{cosec} \theta)^2 = \dfrac{1 + \cos \theta}{1 - \cos \theta}$, where $\cos \theta \neq -1$ [5]

**Q4** (i) By sketching suitable graphs, show that the equation

$$2 - \tfrac{1}{2}x = 1 + \cos x$$

has one root in the interval $0 < x < \pi$ [2]

(ii) Verify, by calculation, that this root lies between 1 and 1.2. [2]

(iii) Use the iterative formula

$$x_{n+1} = \cos^{-1}\left(1 - \tfrac{1}{2}x_n\right)$$

to determine the root correct to 2 decimal places, starting with $x_1 = 1$ and showing each iteration to 4 decimal places. [3]

**Q5** Let $I = \displaystyle\int_0^{\frac{1}{2}} \sqrt{1 - x^2}\, dx$

(i) Using the substitution $x = \sin \theta$, show that

$$I = \int_0^{\frac{\pi}{6}} \cos^2 \theta\, dx$$ [3]

(ii) Hence find the exact value of $I$. [4]

**Q6** (i) (a) The complex numbers $\alpha$ and $\beta$ are given by

$$\alpha = \frac{4}{\sqrt{2} - \sqrt{2}\,i} \text{ and } \beta = -\sqrt{3} + i$$

Show that $\alpha = \sqrt{2} + \sqrt{2}\,i$ and that $|\alpha| = |\beta|$ [3]

(b) Give the exact values of arg $\alpha$ and arg $\beta$. [2]

(ii) Show that $z = \alpha + \beta$ satisfies $|z - \alpha| = |z - \beta|$ [3]

Q7  With respect to the origin, the points $A$, $B$, $C$ have position vectors $\mathbf{a} = -2\mathbf{i} + 2\mathbf{j} + 4\mathbf{k}$, $\mathbf{b} = 4\mathbf{i} + 5\mathbf{j} - 5\mathbf{k}$, $\mathbf{c} = 3\mathbf{i} + 6\mathbf{j} + 4\mathbf{k}$.

   (i) Find a vector equation for the line $AB$. [1]

   (ii) Find the position vector for the point $P$, on $AB$, such that $CP$ is perpendicular to $AB$. [4]

   (iii) Find a Cartesian equation of the plane containing $A$, $B$, $C$. [4]

Q8  Let $f(x) = \dfrac{1 - 2x}{(2 + x)(1 + x^2)}$

   (i) Express $f(x)$ in partial fractions. [5]

   (ii) Hence, assuming that $|x| < 1$, obtain the expansion of $f(x)$ in ascending powers of $x$, up to and including the term in $x^3$. [5]

Q9

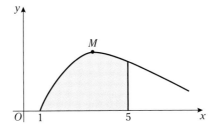

The diagram shows the curve $y = \dfrac{1}{x^2} \ln x$, for $x \geqslant 1$, and its maximum point $M$.

   (i) Find the exact coordinates of $M$. [5]

   (ii) Find the exact area of the shaded region bounded by the curve, the $x$-axis and the line $x = 5$ [5]

Q10  At time $t$ days, the rate of decay of the mass of a radioactive material is proportional to the mass, $m$ grams, of the radioactive material that is present at that time. At $t = 0$, $m = 150$ and at $t = 5$, $m = 100$

   (i) Show that $m = 150e^{-\frac{1}{5}\ln\left(\frac{3}{2}\right)t}$ [7]

   (ii) Find the length of time that it takes for the mass of the radioactive material present to be halved. [3]

## Sample paper 2

**Q1**    (i) Given that $x = 3^y$, show that $3x^2 - 26x - 9 = 3 \times 3^{2y} - 26 \times 3^y - 9$    [1]

    (ii) Hence solve $3 \times 3^{2y} - 26 \times 3^y - 9 = 0$    [4]

**Q2**    Given that the polynomial $\mathrm{p}(x) = 2x^3 + ax^2 + bx - 5$, where $a$ and $b$ are constants to be determined, has a factor $(x - 1)$ and that when $\mathrm{p}(x)$ is divided by $(x + 2)$ the remainder is 27, find $a$ and $b$.    [5]

**Q3**    (i) Show that $x_{r+1} = \frac{1}{2}\left( x_r + \frac{N}{x_r} \right)$ is a possible iterative formula for solving the equation $x^2 = N$    [2]

    (ii) Starting with $x_1 = 4$, and using the iterative formula in part (i), obtain the square root of 19 correct to 4 decimal places. Show the result for each iteration.    [4]

**Q4**    Solve the equation $2 \tan^2 \theta = \sec \theta + 8$ for $0° \leqslant \theta \leqslant 360°$    [7]

**Q5**    (i) A curve has equation $x^3 + 3xy^2 - y^3 = 15$. Show that $\dfrac{dy}{dx} = \dfrac{x^2 + y^2}{y^2 - 2xy}$    [4]

    (ii) Deduce the equation of the tangent to the curve at the point $(2, -1)$.    [3]

**Q6**    Newton's law of cooling may be modelled by the differential equation $\dfrac{dT}{dt} = -k(T - 10)$, where $T$ is the temperature of the body, in °C, at time $t$ minutes, $k$ is a positive constant and room temperature is 10°C.

    Given that when $t = 0$, $T = 100$ and when $t = 10$, $T = 40$, find $T$ in terms of $t$.    [8]

**Q7**    (i) The equation of line $l_1$ is $\mathbf{r} = \mathbf{i} - 2\mathbf{j} + 2\mathbf{k} + \lambda(\mathbf{i} - \mathbf{j} + 2\mathbf{k})$, and the equation of line $l_2$ is $\mathbf{r} = 4\mathbf{i} + \mathbf{j} - 2\mathbf{k} + \mu(\mathbf{i} + 2\mathbf{j} - 3\mathbf{k})$. Find the coordinates of the point of intersection of $l_1$ and $l_2$.    [4]

    (ii) Find a Cartesian equation of the plane containing $l_1$ and $l_2$.    [4]

**Q8**    (i) Obtain the square roots of $2i$ in the form $a + bi$, where $a$ and $b$ are real numbers and $i$ is the square root of $-1$.    [4]

    (ii) Express each of the square roots found in part (i) in the form $re^{i\theta}$, where $r$ is the modulus and $\theta$ is the argument, measured in radians.    [3]

    (iii) Show the position of each square root on an Argand diagram.    [2]

**Q9**    (i) Find the first four non-zero terms of the binomial expansion in ascending powers of $x$ of $(1 - x^2)^{-\frac{1}{2}}$, given that $|x| < 1$    [4]

    (ii) Show that, when $x = \frac{1}{2}$, $(1 - x^2)^{-\frac{1}{2}} = \frac{2}{3}\sqrt{3}$, and hence obtain an approximation to $\sqrt{3}$, giving your answer to 3 significant figures.    [6]

**Q10**    (i) Express $\dfrac{8 - 6x}{(x + 3)(x^2 + 4)}$ in partial fractions.    [5]

    (ii) Hence show that $\displaystyle\int_1^2 \dfrac{8 - 6x}{(x + 3)(x^2 + 4)}\, dx = \ln\left( \dfrac{125}{128} \right)$    [5]

# P2&3 Answers

## Chapter 1

### Exercise 1a

1 quotient $2x + 1$, remainder $-5$
2 quotient $x - 2$, remainder 6
3 quotient $2x^2 + x + 1$, remainder 0
4 quotient $2x^2 + 3x + 6$, remainder 14
5 quotient $x + 1$, remainder $-6x + 3$
6 quotient $x^3 - x^2 + 6x - 6$, remainder 8
7 quotient $3x^2 + 6x + 12$, remainder 19
8 quotient $3x^2 + 3x + 1$, remainder $x + 10$
9 quotient $x^2 + x - 6$, remainder $x + 23$
10 quotient $5x^2 - 14x + 22$, remainder $-10x - 89$

### Exercise 1b

1 (a) 3      (b) 18
  (c) 47      (d) $\frac{35}{16}$
  (e) $-\frac{16}{27}$      (f) $a^3 - 2a^2 + 6$
  (g) $c^2 - ac + b$      (h) $\frac{1}{a^4} - \frac{2}{a} + 1$
2 $-7$
3 yes
4 no
6 neither are
8 (a) $(x - 1)(x + 1)(x + 2)$
  (b) $(x - 2)(x^2 + x + 1)$
  (c) $(2x - 1)(x^2 + 1)$
  (d) $(x + 3)(x - 3)(x^2 + 9)$
  (e) $(x + 3)(x^2 - 3x + 9)$
  (f) $(x - 2)(x + 2)(x^2 + x + 1)$
9 (a) $x = -2$ or $-1$ or $1$    (b) $x = -3$ or $3$
  (c) $x = -3$      (d) $x = -2$ or $2$
10 20
11 $3, -6$
12 (a) 5      (b) $x = 5$
13 $x = -2$ or $2$ or $3$
14 quotient $x^2 - 3x - 3$, remainder 2
15 $a = 4, b = 1$
16 (a) $p = 11, q = -5$    (b) 3
17 (a) $a = 2$      (b)

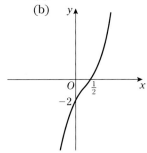

18 $x = 1$ or $x = 2$
19 (a) $a = 5, b = 1$    (b) $(x + 1)^2(3x - 1)$
20 $x^2 - 9$

## Chapter 2

### Exercise 2a

1 $\log_{10} 1000 = 3$
2 $\log_2 16 = 4$
3 $\log_{10} 10\,000 = 4$
4 $\log_3 9 = 2$
5 $\log_4 16 = 2$
6 $\log_5 25 = 2$
7 $\log_{10} 0.01 = -2$
8 $\log_9 3 = \frac{1}{2}$
9 $\log_5 1 = 0$
10 $\log_4 2 = \frac{1}{2}$
11 $\log_{12} 1 = 0$
12 $\log_8 2 = \frac{1}{3}$
13 $\log_q p = 2$
14 $\log_x 2 = y$
15 $\log_p r = q$
16 $10^5 = 100\,000$
17 $4^3 = 64$
18 $10^1 = 10$
19 $2^2 = 4$
20 $2^5 = 32$
21 $10^3 = 1000$
22 $5^0 = 1$
23 $3^2 = 9$
24 $4^2 = 16$
25 $3^3 = 27$
26 $36^{\frac{1}{2}} = 6$
27 $a^0 = 1$
28 $x^z = y$
29 $a^b = 5$
30 $p^r = q$

### Exercise 2b

1 2
2 6
3 6
4 4
5 2
6 3
7 $\frac{1}{2}$
8 $-2$
9 $-1$
10 $\frac{1}{2}$
11 0
12 1
13 $\frac{1}{3}$
14 0

**15** $\frac{1}{3}$

**16** 3

**17** 3

## Exercise 2c

**1** $\log p + \log q$

**2** $\log p + \log q + \log r$

**3** $\log p - \log q$

**4** $\log p + \log q - \log r$

**5** $\log p - \log q - \log r$

**6** $2 \log p + \log q$

**7** $\log q - 2 \log r$

**8** $\log p + \frac{1}{2} \log q$

**9** $2 \log p + 3 \log q - \log r$

**10** $\frac{1}{2} \log q - \frac{1}{2} \log r$

**11** $n \log q$

**12** $n \log p + m \log q$

**13** $\log pq$

**14** $\log p^2 q$

**15** $\log \dfrac{q}{r}$

**16** $\log q^3 p^4$

**17** $\log \dfrac{p^n}{q}$

**18** $\log \dfrac{pq^2}{r^3}$

## Exercise 2d

**1** 2

**2** $-2$

**3** 1.5

**4** 1.63

**5** 1.16

**6** 0.861

**7** 2.77

**8** $\frac{1}{4}$

**9** 1

**10** 16

**11** 1, 4

**12** $x > 3$

**13** $x < 5$

**14** $x > 5$

**15** $x > 2.10$

**16** $x > 1.58$

**17** $\log_x \left(\frac{5}{9}\right), \frac{1}{3}\sqrt{5}$

**18** $\log_3 \left(\dfrac{y}{x^2}\right), y = 3x^2$

**19** $y^2, 1$

**20** $x = 0$

**21** 1.03

## Chapter 3

## Exercise 3a

**1** (a) 7.39      (b) 0.368
    (c) 4.48      (d) 0.741
    (e) 20.1      (f) 6.05
    (g) 0.135      (h) 1.05

**2** (a) $2e^x$      (b) $2x - e^x$
    (c) $e^x$      (d) $-5e^{(2-5x)}$
    (e) $6e^{(2x-4)}$      (f) $-2e^{(2-x)}$
    (g) $4xe^{2x^2}$      (h) $6x + 2xe^{(x^2)}$

**3** $e^2 - 2$

**4** $2 + 2e$

**5** 1

**6** $\ln 3$

**7** (a)

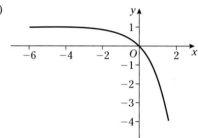

(b)

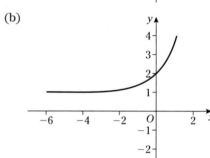

(c)

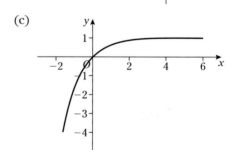

(d)

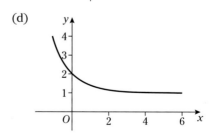

## Exercise 3b

1   (a)  $\ln 4 = x$     (b)  $\ln y = 2$     (c)  $\ln b = a$
2   (a)  $e^4 = x$        (b)  $e^a = 0.5$     (c)  $e^b = a$
3   (a)  1.10             (b)  0.875           (c)  −1.60
    (d)  2.85
4   (a)  1                (b)  2               (c)  0
5   (a)  $\ln 5 + \ln x$
    (b)  $\ln 5 + 2 \ln x$
    (c)  $\ln 3 + \ln (x + 1)$
    (d)  $\ln (x + 1) - \ln x$
    (e)  $\ln (2x - 1) - \ln x$
    (f)  $\ln x + 2 \ln y$
    (g)  $\frac{1}{2} \ln (x + 1)$
    (h)  $\ln x + \ln (x + 4)$
    (i)  $\ln (x + 1) + \ln (x - 1)$
    (j)  $2 \ln x + \ln (x + y)$
    (k)  $1 + \ln x$
    (l)  $2 + \ln x + \ln (x - e)$
    (m)  $2 \ln x - \ln (x + 1)$
    (n)  $\ln (a + b) + \ln (a - b)$
    (o)  $\ln \sin x - \ln \cos x$
6   (a)  $\ln 2x$                    (b)  $\ln \left(\dfrac{3}{x}\right)$
    (c)  $\ln \left(\dfrac{x^2}{4}\right)$     (d)  $\ln \left(\dfrac{x}{(1 - x)^2}\right)$
    (e)  $\ln \left(\dfrac{e}{x}\right)$       (f)  $\ln (e^2 x)$
    (g)  $\ln \left(\dfrac{x^2}{\sqrt{x - 1}}\right)$   (h)  $\ln \dfrac{\cos x}{\sin x}$
    (i)  $\ln (ex)$                   (j)  $\ln (x - 1)^{\frac{2}{3}}$
7   (a)  2.10            (b)  0              (c)  1.05
    (d)  $\frac{3}{2}$ or $-1$    (e)  $-3$ or $1$
8   $\ln (e + 1)$
9   $\dfrac{3e^3}{(e^3 - 1)}$

## Exercise 3c

1   (a)  $\dfrac{3}{x}$     (b)  $\dfrac{1}{x}$     (c)  $-\dfrac{2}{x}$     (d)  $-\dfrac{1}{2x}$
    (e)  $-\dfrac{5}{x}$    (f)  $\dfrac{1}{2x}$    (g)  $-\dfrac{3}{2x}$    (h)  $\dfrac{5}{2x}$
2   (a)  $(1, -1)$
    (b)  $\left(2^{\frac{1}{3}}, \{2 - 2 \ln 2\}\right)$
    (c)  $(4, \{\ln 4 - 2\})$
3   (a)

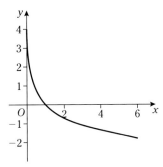

(b)

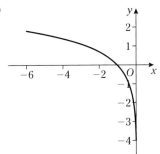

(c)

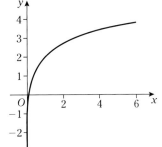

(d)

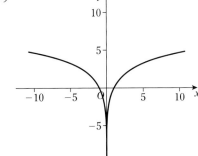

4   (a)  $\dfrac{3}{3x - 1}$              (b)  $\dfrac{4}{4x + 3}$
    (c)  $\dfrac{2x}{x^2 + 1}$            (d)  $\dfrac{2x + 2}{x^2 + 2x}$
    (e)  $\dfrac{3x^2}{x^3 + 2}$

## Mixed exercise 3d

1   $y = \dfrac{x}{1 - x}$
2

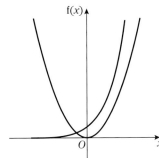

f(x)    there is only one point of intersection

3   1
4   $y - 2 \ln 4 = \frac{1}{2} (x - 4)$

**5** $y - 3e^2 = -\dfrac{1}{6e^2}(x - 1)$

**6** $\dfrac{dy}{dx} = 2e^{2x} - \dfrac{1}{2x}$

**7** (a) $3e^{3x+2}$  (b) $\dfrac{2x}{x^2 - 3}$  (c) $2xe^{x^2-1}$

    (d) $\dfrac{2x}{x^2 - 1}$  (e) $\dfrac{3x^2 + 2}{x^3 + 2x + 1}$

## Exercise 3e

**1** $a = 0.091, b = 0.787$

**2** $a = 148, b = 0.607$

**3** (b) $b = 1$

**4** 2

**5** $(1.83, 0)$

**6** $a = 8760, b = 1.13$

**7** $\dfrac{1}{25}$

**8** (a) $\ln s = \ln k - nt$: plotting $\ln s$ against $t$ gives a straight line

    (b) $k = 5500, n = 1.5$

## Chapter 4

## Exercise 4a

**1**
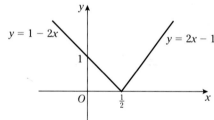

**2**
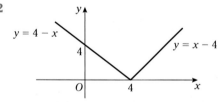

**3**
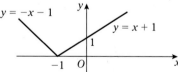

**4**
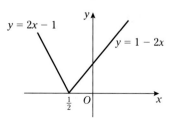

**5**
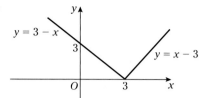

**6**
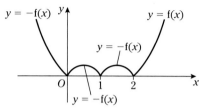

$f(x) = x(x - 1)(x - 2)$

**7**
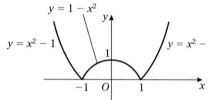

**8**
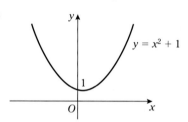

**9**
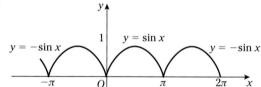

**10**
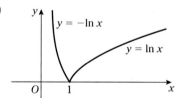

**11**
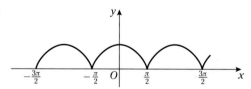

**12**

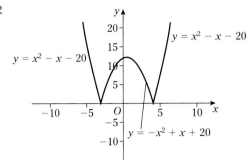

$y = x^2 - x - 20$

$y = x^2 - x - 20$

$y = -x^2 + x + 20$

## Exercise 4b

1  $(-1, 2)$, $\left(\frac{1}{3}, \frac{2}{3}\right)$

2  $\left(\frac{1}{3}, \frac{1}{3}\right)$, $(1, 1)$

3  $(-2, -1)$, $\left(-\frac{6}{5}, -\frac{3}{5}\right)$

4  $(5, 5)$, $\left(\frac{5}{3}, \frac{5}{3}\right)$

5  $(1, 3)$

6  $\frac{1}{2}$

7  $-\frac{2}{3}$, $2$

8  $-4, 8$

9  $3$ and $\frac{1}{2}\left(\sqrt{17} - 3\right)$

10  $\frac{4}{3}$ and $\frac{2}{5}$

11  $2$ and $\frac{2}{3}$

12  $-1$ and $-\sqrt{2} - 1$

13  $\frac{1}{2}$

14  $2$ and $3$

15  $\ln 4$

## Exercise 4c

1  $x < -2, x > 4$

2  $x > -3, x < 2$

3  $x < -\frac{1}{2}, x > 1$

4  $x > -\frac{1}{2}$

5  $x < 0, x > 2$

6  $x < 1, x > 3$

7  $x > 1$

8  $-1 < x < 1$

9  $-2 < x < \frac{2}{3}$

10  $x < -\frac{1}{2}, x > \frac{1}{2}$

11  $x < -1, x > \frac{1}{2}$

12  $x < 1, x > \frac{5}{3}$

13  $x < 1$

14  $x < 0$

15  $0 < x < 1 + \sqrt{3}$

16  $0 < x < 1$

17  $x > 1$

18  $x < -1 + \sqrt{6}, x > 5$

## Chapter 5

## Exercise 5a

1  $(x - 3)^2 + 2x(x - 3) = 3(x - 3)(x - 1)$

2  $\sqrt{x - 6} + \dfrac{x}{2\sqrt{x - 6}} = \dfrac{3(x - 4)}{2\sqrt{x - 6}}$

3  $e^{2x} + xe^{2x}$

4  $(2x + 3)^3 + 6x(2x + 3)^2 = (2x + 3)^2(8x + 3)$

5  $1 + \ln x$

6  $xe^x$

7  $\dfrac{x + 1}{x} + \ln 2x$

8  $\sqrt{x} + \dfrac{x + 1}{2\sqrt{x}}$

9  $2xe^{x^2} + 2x^3e^{x^2}$

10  $3x^2(x - 1)^2 + 2x^3(x - 1) = x^2(x - 1)(5x - 3)$

11  $(x + 3)^{-1} - x(x + 3)^{-2} = 3(x + 3)^{-2}$

12  $3x^2 \ln(x - 1) + \dfrac{x^3}{x - 1}$

13  $e^x(x^2 + 2x - 1)$

14  $\ln(x^2 - 3x) + \dfrac{2x^2 - 3x}{x^2 - 3x}$

## Exercise 5b

1  $\dfrac{2x(x - 3) - (x - 3)^2}{x^2} = \dfrac{x^2 - 9}{x^2}$

2  $\dfrac{2x(x + 3) - x^2}{(x + 3)^2} = \dfrac{x(x + 6)}{(x + 3)^2}$

3  $\dfrac{x^2 - 8x}{x^4} = \dfrac{x - 8}{x^3}$

4  $\dfrac{x^2(x + 1)(-x - 3)}{x^6} = \dfrac{-(x + 1)(x + 3)}{x^4}$

5  $\dfrac{4(1 + 2x)}{(1 - x)^4}$

6  $\dfrac{2x(x - 4)}{(x - 2)^2}$

7  $\dfrac{(x - 1)e^x - e^x}{(x - 1)^2} = \dfrac{e^x(x - 2)}{(x - 1)^2}$

8  $\dfrac{1 - \ln x}{x^2}$

9  $\dfrac{2e^{2x}(x + 1)^2 - 2e^{2x}(x + 1)}{(x + 1)^4} = \dfrac{2xe^{2x}(x + 1)}{(x + 1)^4}$

10  $\dfrac{1 - x}{e^x}$

11  $\dfrac{x^2e^x - 2xe^x}{x^4} = \dfrac{e^x(x - 2)}{x^3}$

12  $\dfrac{x^2 - 3x^2 \ln x}{x^6} = \dfrac{1 - 3 \ln x}{x^4}$

13  $\dfrac{(x^2 - 1)e^x - 2xe^x}{(x^2 - 1)^2}$

14  $\dfrac{-2}{(e^x - e^{-x})^2}$

15  $-\dfrac{1}{x (\ln x)^2}$

## Mixed exercise 5

1  $\dfrac{3x + 2}{2\sqrt{x + 1}}$

2  $6x(x^2 - 8)^2$

3  $\dfrac{1 - x^2}{(x^2 + 1)^2}$

4  $\dfrac{-4x^3}{3(2 - x^4)^{\frac{2}{3}}}$

5  $\dfrac{2x}{(x^2 + 2)^2}$

6  $\frac{1}{2}x(5\sqrt{x} - 8)$

7  $6x(x^2 - 2)^2$

8  $\dfrac{1 - 2x}{2\sqrt{x - x^2}}$

9  $\dfrac{\sqrt{x} + 2}{2(\sqrt{x} + 1)^2}$

10  $\dfrac{x(5x - 8)}{2\sqrt{x - 2}}$

11  $\dfrac{-(3x + 4)}{2x^3\sqrt{x + 1}}$

12  $6x^5(x^2 + 1)^2(2x^2 + 1)$

13  $\dfrac{x}{\sqrt{x^2 - 8}}$

14  $x^2(5x^2 - 18)$

15  $6x(x^2 - 6)^2$

16  $\dfrac{-(x^2 + 6)}{(x^2 - 6)^2}$

17  $-8x^3(x^4 + 3)^{-3}$

18  $\dfrac{(2 - x)^2(2 - 7x)}{2\sqrt{x}}$

19  $\dfrac{2 + 5x}{2\sqrt{x}(2 - x)^4}$

20  $(x - 2)(3x - 4)$

21  $30x^2(2x^3 + 4)^4$

22  $1 + \ln x$

23  $\frac{8}{3}(4x - 1)^{-\frac{1}{3}}$

24  $\dfrac{e^x(x - 2)}{(x - 1)^2}$

25  $\dfrac{-(x^3 + 4)}{2x^3\sqrt{1 + x^3}}$

26  $\dfrac{(x - 1)\ln(x - 1) - x\ln x}{x(x - 1)\{\ln(x - 1)\}^2}$

27  $\dfrac{2\ln x}{x}$

28  $\dfrac{2x}{(1 + x^2)^2}$

29  $\dfrac{2}{x^2 e^{\frac{2}{x}}}$

30  $\dfrac{-e^x}{1 - e^x}$ or $\dfrac{e^x}{e^x - 1}$

31  $3x^2 e^{3x}(x + 1)$

32  $\dfrac{4}{5(2x - 1)^2} - \dfrac{6}{5(x - 3)^2} = \dfrac{2(3 - 2x^2)}{(2x - 1)^2(x - 3)^2}$

33  $\dfrac{e^{\frac{x}{2}}(x - 10)}{2x^6}$

34  $\dfrac{2}{x} - \dfrac{1}{x + 3}$

35  $\dfrac{5x + 9}{x(x + 3)}$

36  $\dfrac{4}{x}(\ln x)^3$

37  $\dfrac{(x + 3)^2(x^2 - 6x + 6)}{(x^2 + 2)^2}$

38  $\dfrac{e^x - 1}{2\sqrt{e^x - x}}$

39  $\dfrac{8x}{x^2 + 1}$

40  $\dfrac{dy}{dx} = \dfrac{4}{(1 - 2x)^2}; \dfrac{d^2y}{dx^2} = \dfrac{16}{(1 - 2x)^3}$

41  $\dfrac{dy}{dx} = \dfrac{1}{x(x + 1)}; \dfrac{d^2y}{dx^2} = -\dfrac{(2x + 1)}{x^2(x + 1)^2}$

42  $\dfrac{dy}{dx} = \dfrac{-4e^x}{(e^x - 4)^2}; \dfrac{d^2y}{dx^2} = \dfrac{4e^x(e^x + 4)}{(e^x - 4)^3}$

## Summary exercise 1

1  $x < \frac{3}{2}$

2  2.32

3  (i)  $-1$   (ii)  $x = -2, -\frac{1}{2},$ or 3

4  $x < \frac{1}{2}$

5  (i)  $\dfrac{\ln 4}{\ln 3}$   (ii)  3.42

6  $-\frac{1}{3} < x < 1$

7  2.41

8  $a = 2, b = -3$

9  4.11

10  $-1 < x < -\frac{1}{2}$

11  $a = -4, b = 1$

12  (i)  $y^2$   (ii)  1 or $\dfrac{\ln 0.5}{\ln 3}$

13  2.49

14  $-1 < x < 5$

15  (i)  $4 < y < 6$   (ii)  $1.26 < x < 1.63$

16  (i)  $x = 2$

17  $A = 3.67, b = 1.28$

18  $(4, 4e^{-2})$

19  (i)  $\left(\dfrac{1}{e}, -\dfrac{1}{e}\right)$   (ii)  minimum

20  $x = 1.22$

21  $k \approx 10, a = 2.2$

# Chapter 6

## Exercise 6a

1. (a) $60°, 300°$     (b) $59.0°, 239.0°$
   (c) $41.8°, 138.2°$
2. (a) $-140.2°, 39.8°$     (b) $-131.8°, 131.8°$
   (c) $-150°, -30°$
3. $-\frac{1}{2}\pi, \frac{1}{2}\pi$
4. (a) $1$     (b) $-\sqrt{2}$
   (c) $-2$

## Exercise 6b

1. $\tan^4 A$
2. $\sec^2 \theta$
3. $\tan \theta$
4. $\sin^3 \theta$
5. $x^2 - y^2 = 16$
6. $x^2(b^2 - y^2) = a^2 b^2$
7. $x^2(b^2 - y^2) = a^2 b^2$
11. $38.2°, 141.8°$
12. $57.7°, 122.3°, 237.7°, 302.3°$
13. $30°, 150°$
14. $30°, 150°$
15. $45°, 166.2°, 225°, 346°$
16. $199°, 341°$

## Exercise 6d

1. $0$
2. $\frac{1}{2}$
3. $\frac{1}{4}(\sqrt{6} - \sqrt{2})$
4. $-(2 + \sqrt{3})$
5. $\frac{1}{4}(\sqrt{6} - \sqrt{2})$
6. $\frac{1}{4}(\sqrt{6} + \sqrt{2})$
7. $\sin 3\theta$
8. $0$
9. $\tan 3A$
10. $\tan \beta$
16. $67.5°, 247.5°$
17. $7.4°, 187.4°$
18. $37.9°, 217.9°$
19. $15°, 195°$
20. $0, 60°, 90°, 120°, 180°, 240°, 270°, 300°, 360°$

## Exercise 6e

1. $\frac{1}{2}$
2. $\frac{1}{\sqrt{2}}$
3. $\frac{1}{2} \sin 2\theta$
4. $\cos 8\theta$
5. $-\frac{1}{\sqrt{3}}$
6. $\tan 6\theta$

7. $-\dfrac{1}{\sqrt{2}}$

8. $\dfrac{1}{\sqrt{2}}$

9. (a) $-\frac{7}{25}, \frac{24}{25}$     (b) $\frac{527}{625}, \frac{336}{625}$
   (c) $-\frac{119}{169}, \frac{120}{169}$

10. (a) $-\frac{336}{527}$     (b) $\frac{527}{625}$
    (c) $-\frac{336}{625}$     (d) $\frac{164\,833}{390\,625}$

11. (a) $x(1 - y^2) = 2y$     (b) $x = 2y^2 - 1$
    (c) $x = 1 - \dfrac{2}{y^2}$     (d) $2x^2 y + 1 = y$

12. (a) $-\cos 2x$
    (b) $3 - \cos 2x$
    (c) $\frac{1}{2}(\cos 2x + 3)$
    (d) $\frac{1}{2}(\cos 2x + 1)(3 + \cos 2x)$
    (e) $\frac{1}{4}(1 + \cos 2x)^2$
    (f) $\frac{1}{4}(1 - \cos 2x)^2$

14. (a) $\frac{1}{6}\pi, \frac{5}{6}\pi, \frac{3}{2}\pi$     (b) $\frac{1}{2}\pi, \frac{7}{6}\pi, \frac{3}{2}\pi, \frac{11}{6}\pi$
    (c) $0, \frac{2}{3}\pi, \frac{4}{3}\pi, 2\pi$     (d) $\frac{1}{6}\pi, \frac{1}{2}\pi, \frac{5}{6}\pi, \frac{3}{2}\pi$
    (e) $\frac{1}{3}\pi, \frac{5}{3}\pi$     (f) $\frac{1}{4}\pi, \frac{1}{2}\pi, \frac{5}{4}\pi, \frac{3}{2}\pi$

## Exercise 6f

1. (a) $2, 30°$     (b) $\sqrt{10}, 71.6°$
   (c) $5, 36.9°$
2. $\sqrt{2} \cos\left(2\theta + \frac{1}{4}\pi\right)$
3. $\sqrt{29} \sin(3\theta + 21.8°)$
4. $-2 \sin\left(\theta - \frac{1}{6}\pi\right)$; max $2$ at $\theta = 300°$,
   min $-2$ at $\theta = 120°$
5. $25 \cos(\theta + 73.7°)$; max $28$ at $\theta = 286.3°$,
   min $-22$ at $\theta = 106.3°$
6. $\sqrt{2}, -\sqrt{2}; -\dfrac{1}{\sqrt{2}}$ max, $\dfrac{1}{\sqrt{2}}$ min
7. (a) $45°$     (b) $118.1°, 323.1°$
   (c) $0, 216.80°, 360°$     (d) $0, 306.9°, 360°$

## Mixed exercise 6

1. $y = 1 - 2x^2$
4. $\frac{56}{65}, -\frac{16}{65}$
5. $x = 2y - 1$
7. $5 \sin(\theta - \alpha)$ where $\tan \alpha = \frac{3}{4}, 7, -3$
8. $\sqrt{2} \sin(2\theta - 45°), 67.5°$
9. $-\pi, 0, \pi$
11. $\cot^2 x$
12. $90°, 270°$
13. (a) $2 - \cos 2\theta$     (b) $2 + 2\cos 4A$
14. $0$
15. $40.2°$

# Chapter 7

## Exercise 7a

1. (a) $\cos x + \sin x$      (b) $\cos \theta$
   (c) $-3 \sin \theta$      (d) $5 \cos \theta$
   (e) $3 \cos \theta - 2 \sin \theta$      (f) $4 \cos x + 6 \sin x$
2. (a) $-1$      (b) $1$      (c) $-1$
   (d) $1$      (e) $2(\pi - 1)$      (f) $4$
3. (a) $\frac{1}{6}\pi$      (b) $\frac{1}{6}\pi$
   (c) $\frac{1}{4}\pi$      (d) $\pi$
4. (a) $\left(\frac{1}{3}\pi, \sqrt{3} - \frac{1}{3}\pi\right)$, max; $\left(\frac{5}{3}\pi, -\sqrt{3} - \frac{5}{3}\pi\right)$, min
   (b) $\left(\frac{1}{6}\pi, \frac{1}{6}\pi + \sqrt{3}\right)$, max; $\left(\frac{5}{6}\pi, \frac{5}{6}\pi - \sqrt{3}\right)$, min
5. $y + \theta = 3 + \frac{1}{2}\pi$
6. $2\pi y + x = 2\pi^3 - \pi$
7. $(0, 1)$

## Exercise 7b

1. $4 \cos 4x$
2. $2 \sin(\pi - 2x)$ or $2 \sin 2x$
3. $\frac{1}{2}\cos\left(\frac{1}{2}x + \pi\right)$ or $-\frac{1}{2}\cos\frac{1}{2}\pi$
4. $\dfrac{x \cos x - \sin x}{x^2}$
5. $-\dfrac{(\cos x + \sin x)}{e^x}$
6. $\dfrac{\cos x}{2\sqrt{\sin x}}$
7. $2 \sin x \cos x$ or $\sin 2x$
8. $\cos^2 x - \sin^2 x$ or $\cos 2x$
9. $\cos x\, e^{\sin x}$
10. $-\tan x$
11. $e^x(\cos x - \sin x)$
12. $x^2 \cos x + 2x \sin x$
13. $\dfrac{\sin x}{\cos^2 x} = \sec x \tan x$
14. $\sec^2 x$
15. $\dfrac{-\cos x}{\sin^2 x} = -\cosec x \cot x$
16. $\dfrac{-1}{\sin^2 x} = -\cosec^2 x$

## Mixed exercise 7

1. (a) $-4 \cos 4\theta$
   (b) $1 + \sin \theta$
   (c) $3 \sin^2 \theta \cos \theta + 3 \cos 3\theta$
2. (a) $3x^2 + e^x$
   (b) $2e^{(2x + 3)}$
   (c) $e^x(\sin x + \cos x)$
3. (a) $3 \cos x + e^{-x}$
   (b) $4x^3 + 4e^x - \dfrac{1}{x}$
4. $1 + \dfrac{1}{x} + \ln x$
5. $3 \sin 6x$

6. $\frac{8}{3}(4x - 1)^{-\frac{1}{3}}$
7. $\dfrac{(x^4 + 4x^3 + 3)}{(x + 1)^4}$
8. $2x \sin x + x^2 \cos x$
9. $\dfrac{e^x(x - 2)}{(x - 1)^2}$
10. $\dfrac{2 \cos x}{(1 - \sin x)^2}$
11. $\dfrac{x(5x - 4)}{2\sqrt{x - 1}}$
12. $\cos^2 x(4 \cos^2 x - 3)$
13. (a) $1$      (b) $y - x = 1 - \frac{1}{2}\pi$
    (c) $y + x = 1 + \frac{1}{2}\pi$
14. (a) $1 + e$      (b) $y = x(1 + e)$
    (c) $y(1 + e) + x = (1 + e)^2 + 1$
15. (a) $2$      (b) $y = 2x + 1$
    (c) $2y + x = 2$
16. (a) $-1$      (b) $x + y = 3$
    (c) $x - y + 1 = 0$
17. (a) $\left(\frac{1}{2}\pi, 0\right)$, min; $\left(\frac{3}{2}\pi, 2\right)$, max
    (b) $\left(\frac{1}{6}\pi, \left\{\frac{1}{12}\pi + \frac{1}{2}\sqrt{3}\right\}\right)$, max;
    $\left(\frac{5}{6}\pi, \left\{\frac{5}{12}\pi - \frac{1}{2}\sqrt{3}\right\}\right)$, min
18. (a) $\left(\frac{1}{6}\pi, \left\{\frac{1}{2}\pi - \sqrt{3}\right\}\right)$      (b) $(1, -1)$

# Chapter 8

## Exercise 8a

1. $2x + 2y\dfrac{dy}{dx} = 0$
2. $2x + y + (x + 2y)\dfrac{dy}{dx} = 0$
3. $2x + x\dfrac{dy}{dx} + y = 2y\dfrac{dy}{dx}$
4. $-\dfrac{1}{x^2} - \dfrac{1}{y^2}\dfrac{dy}{dx} = e^y\dfrac{dy}{dx}$
5. $\cos x + \cos y\dfrac{dy}{dx} = 0$
6. $e^y + xe^y\dfrac{dy}{dx} = 1$
7. $\dfrac{dy}{dx} = \pm\dfrac{1}{\sqrt{2x + 1}}$
8. $\pm\frac{1}{4}\sqrt{2}$
9. $3x + 12y - 7 = 0$
10. (a) $\dfrac{dy}{dx} = \dfrac{\sin x}{\cos y}$
    (b) $\dfrac{\pi}{4}$
11. (a) $\dfrac{dy}{dx} = -\dfrac{x^2}{y^2}$      (b) $y = 3, 9y = -x + 28$
12. (a) $\dfrac{dy}{dx} = -\dfrac{x}{y}$      (b) $y = 2a\sqrt{2} - x$
13. $y = 2x + 1$ or $y = 2x - 1$

# Exercise 8b

1  (a) $\dfrac{1}{4t}$  (b) $-\cot\theta$  (c) $-\dfrac{4}{t^2}$

2  $\dfrac{dy}{dx}=2t-t^2;\dfrac{3}{4}$

3  $\dfrac{3}{2}t$

4  $\dfrac{dy}{dx}=e^{2t}$

5  $\left(-\dfrac{1}{3}\sqrt3,\dfrac{2}{9}\sqrt3\right)$, max; $\left(\dfrac{1}{3}\sqrt3,-\dfrac{2}{9}\sqrt3\right)$, min

6  $\dfrac{1}{2}\pi$

7  $2x+y+2=0$

8  $6y=4x+5\sqrt2,\left(-\dfrac{137}{97}\sqrt2,-\dfrac{21}{194}\sqrt2\right)$

# Mixed exercise 8

1  (a) $4y^3\dfrac{dy}{dx}$  (b) $y^2+2xy\dfrac{dy}{dx}$

(c) $-\dfrac{1}{y^2}\dfrac{dy}{dx}$  (d) $\ln y+\dfrac{x}{y}\dfrac{dy}{dx}$

(e) $\cos y\dfrac{dy}{dx}$  (f) $e^y\dfrac{dy}{dx}$

(g) $\dfrac{dy}{dx}\cos x-y\sin x$

(h) $(\cos y-y\sin y)\dfrac{dy}{dx}$

2  $\dfrac{x}{2y}$

3  $-\dfrac{y^2}{x^2}$

4  $-\dfrac{2y}{3x}$

5  $\dfrac{3t}{2}$

6  $\dfrac{t}{t+1}$

7  $-\dfrac{3}{2}\cos\theta$

8  $-\dfrac{1}{t^2}$

9  $-\dfrac{1}{e^t}$

10  $2t-t^2$

11  $\dfrac{dy}{dx}=-4x$

13  $2y\dfrac{dy}{dx}-2x\dfrac{dy}{dx}-2y+3\dfrac{dy}{dx}=7$
   $37y=81x+1$

14  $4y+2x=3\sqrt2$

# Chapter 9

## Exercise 9a

Add a constant to each answer.

1  $\dfrac{1}{4}e^{4x}$

2  $-4e^{-x}$

3  $\dfrac{1}{3}e^{(3x-2)}$

4  $-\dfrac{2}{5}e^{(1-5x)}$

5  $-3e^{-2x}$

6  $5e^{(x-3)}$

7  $2e^{(\frac{x}{2}+2)}$

8  $\dfrac{1}{2}e^{2x}-\dfrac{1}{2e^{2x}}$

9  $\dfrac{e^4-1}{2}$

10  $2(e^2-1)$

11  $\dfrac{e-1}{e}$

12  $1-e^2$

## Exercise 9b

1  $\dfrac{1}{2}\ln|x|+K$

2  $2\ln|x|+K$

3  $\dfrac{1}{4}\ln|x|+K$

4  $\dfrac{3}{2}\ln|x|+K$

5  $4\ln|x-1|+K$

6  $\dfrac{1}{3}\ln|3x+1|+K$

7  $-\dfrac{3}{2}\ln|1-2x|+K$

8  $2\ln|2+3x|+K$

9  $-\dfrac{3}{2}\ln|4-2x|+K$

10  $-4\ln|1-x|+K$

11  $-\dfrac{5}{7}\ln|6-7x|+K$

12  $3\ln1.5$

13  $\dfrac{1}{2}\ln3$

14  $2\ln2=\ln4$

15  $\ln2$

## Exercise 9c

1  $-\dfrac{1}{2}\cos2x+K$

2  $\dfrac{1}{7}\sin7x+K$

3  $\dfrac{1}{4}\tan4x+K$

4  $-\cos\left(\dfrac{1}{4}\pi+x\right)+K$

5  $\dfrac{3}{4}\sin\left(4x-\dfrac{1}{2}\pi\right)+K$

6  $\dfrac{1}{2}\tan\left(\dfrac{\pi}{3}+2x\right)+K$

7  $-\dfrac{3}{5}\cos5x+K$

8  $-\dfrac{2}{3}\cos(3x-\alpha)+K$

9  $-10\sin\left(\alpha-\dfrac{1}{2}x\right)+K$

10  $-\left(\dfrac{1}{4}\right)\cos(4x-\pi)+K$

11  $\dfrac{1}{3}\sin3x-\sin x+K$

12  $\dfrac{1}{2}\tan2x+K$

13  $\dfrac{1}{3}$

14  $-\dfrac{1}{4}$

15  $0$

16  $\dfrac{1}{2}$

17  $-\ln\left(\dfrac{1}{2}\right)$

## Exercise 9d

1   $-\frac{1}{3}\cos 3x + K$

2   $\frac{1}{2}$

3   $\frac{1}{4}\sin 4x + K$

4   (a)   $2 - \cos 2x$      (b)   $\frac{x}{2} - \frac{\sin 2x}{4} + K$

5   $-x + \frac{1}{2}\sin 2x - \cos 2x + K$

6   $\pi - 4$

7   $2\sin^3 x + K$

## Exercise 9e

1   0.806

2   1.24

3   1.462

4   0.529

5   1 is an overestimate, 2 is an underestimate

6   (a)   11.2      (b)   overestimate

7   (a)   0.512      (b)   underestimate

## Mixed exercise 9

1   $\frac{3}{2}e^{2x-1} + K$

2   $\frac{1}{3}\ln|x| + K$

3   $\frac{1}{2}\sin(2x + \pi)$

4   $2\ln|1 + x| + K$

5   $\frac{1}{2}\tan(2x - 1) + K$

6   $e^x - e^{-x}$

7   $\frac{1}{3}\cos\left(\frac{\pi}{3} - 3x\right) + K$

8   $\frac{4}{3}\ln|3x - 2| + K$

9   $\frac{3}{4}(e^3 - e^{-1})$

10   $\frac{1}{4}$

11   $\ln 2$

12   $\frac{1}{\sqrt{2}}$

13   $\frac{3\pi}{8} - \frac{1}{4}$

14   $\frac{8}{15}$

## Chapter 10

## Exercise 10a

1   (a)   infinite    (b)   2       (c)   2
     (d)   1          (e)   3       (f)   2
     (g)   1          (h)   1       (i)   1

2   (b)   $1 < x < 1.5$      (c)   0 (exact)
     (d)   0 (exact)           (e)   $2 < x < 2.5$
     (f)   $0.5 < x < 1$      (g)   $1.5 < x < 2$
     (h)   $0.5 < x < 1$      (i)   $0 < x < 0.5$

3   $(0, 1)$ max, $(2, -3)$ min, 3

4   (a)   2       (b)   3       (c)   3

5   $-1$

6   $-3 < x < -2$

## Exercise 10b

5   2.0801, 2.0984, 2.1026

6   0.20

8   (a)   2.0       (b)   $3x^2 = 4x + 4$, 2

9   (a)   3.15      (b)   $x = \ln x + 2$

10   0.16

11   any rearrangement of the equations in the form $x = f(x)$

## Summary exercise 2

1   (i)   3.1555, 3.1416..., 3.1415...; $\alpha = 3.1412$
    (ii)   $5x = 4x + \frac{306}{x^4}$

2   (i)   $5\sin(\theta + \alpha)$, $\alpha = 53.13°$
    (ii)   $11.0°$, $62.7°$
    (iii)   2

3   (i)   $\frac{dy}{dx} = \frac{2t(t-1)}{3t-2}$      (ii)   $(6, 5)$

4   (ii)   $(2, 1)$, $(-2, 1)$

5   (i)   $(0, 1)$          (ii)   $\frac{\pi}{4}$
    (iii)   1.77         (iv)   underestimate

6   (i)   $\sqrt{26}\cos(\theta + 11.31)$
    (ii)   $27.0°$, $310.4°$

7   $48.2°$, $120°$

9   (i)   1.21         (ii)   overestimate

10   (iv)   0.27

11   (ii)   $73.9°$, $253.9°$      (iii)   $-\frac{11}{13}$

12   (ii)   $(1, 2)$, $(-1, -2)$

13   (iii)   $\sqrt{3} - \frac{\pi}{3}$      (iv)   $\frac{\sqrt{3}}{2}$

14   (i)   2        (iii)   0.95

15   $\frac{1}{4}(5\pi - 2)$

16   (i)   $(1, 1)$, $(1, -3)$      (ii)   $2x + y + 1 = 0$

17   (ii)   $10.9°$, $-169.1°$

18   (i)   $\left(\frac{1}{e}, -\frac{1}{e}\right)$      (ii)   minimum

## Chapter 11

## Exercise 11a

1   $\frac{3}{2(x+1)} - \frac{1}{2(x-1)}$

2   $\frac{13}{6(x-7)} - \frac{1}{6(x-1)}$

3   $\frac{4}{5(x-2)} - \frac{4}{5(x+3)}$

4   $\frac{7}{9(2x-1)} + \frac{28}{9(x+4)}$

5   $\frac{1}{x-2} - \frac{1}{x}$

6   $\frac{3}{x-2} - \frac{1}{x-1}$

7   $\frac{1}{2(x-3)} - \frac{1}{2(x+3)}$

**8** $\dfrac{7}{3x} - \dfrac{1}{3(x + 1)}$

**9** $\dfrac{9}{x} - \dfrac{18}{2x + 1}$

**10** $\dfrac{2}{5(x - 1)} - \dfrac{1}{5(3x + 2)}$

**11** $\dfrac{1}{x - 1} - \dfrac{1}{x + 1}$

**12** $\dfrac{1}{(x - 2)} - \dfrac{1}{(x + 1)}$

**13** $\dfrac{1}{3(x - 3)} - \dfrac{1}{3x}$

**14** $\dfrac{1}{x - 1} - \dfrac{1}{x + 3}$

**15** $\dfrac{1}{2(x - 1)} - \dfrac{1}{2(x + 1)}$

**16** $\dfrac{1}{2x - 1} - \dfrac{1}{2x + 1}$

**17** $\dfrac{1}{x - 1} + \dfrac{1}{x + 1} + \dfrac{1}{x + 2}$

**18** $\dfrac{1}{2(x + 3)} - \dfrac{1}{4(x - 1)} + \dfrac{3}{4(x + 2)}$

## Exercise 11b

**1** $\dfrac{1}{2(x - 1)} - \dfrac{1}{2(x + 1)} - \dfrac{1}{(x + 1)^2}$

**2** $\dfrac{3}{2x} - \dfrac{x}{2(x^2 + 2)}$

**3** $\dfrac{11}{8(x - 3)} + \dfrac{5}{8(x + 1)} - \dfrac{1}{2(x + 1)^2}$

**4** $\dfrac{1}{x} - \dfrac{x}{2x^2 + 1}$

**5** $\dfrac{1}{x - 1} - \dfrac{1}{x - 2} + \dfrac{2}{(x - 2)^2}$

**6** $\dfrac{2}{x} - \dfrac{1}{x^2} - \dfrac{3}{2x + 1}$

**7** $\dfrac{7}{16(x + 3)} - \dfrac{1}{4(x - 1)^2} + \dfrac{9}{16(x - 1)}$

**8** $\dfrac{3}{(x + 2)^2} + \dfrac{2}{x + 2} - \dfrac{2}{x + 1}$

**9** $\dfrac{1}{2(x - 2)} - \dfrac{1}{6(x + 2)} - \dfrac{1}{3(x - 1)};$
using (b) goes directly to three partial fractions

## Exercise 11c

**1** $1 + \dfrac{1}{2(x - 1)} - \dfrac{1}{2(x + 1)}$

**2** $1 + \dfrac{2}{x - 1} - \dfrac{2}{x + 1}$ or $1 + \dfrac{4}{x^2 - 1}$

**3** $1 - \dfrac{7}{4(x + 3)} - \dfrac{1}{4(x - 1)}$

**4** $1 - \dfrac{2x + 1}{5(x^2 + 1)} - \dfrac{8}{5(x + 2)}$

## Exercise 11d

**1** (a) $\dfrac{4}{7(x - 3)} - \dfrac{8}{7(2x + 1)}$

  (b) $\dfrac{5}{7(x + 1)} + \dfrac{1}{7(4x - 3)}$

  (c) $\dfrac{1}{t - 1} + \dfrac{1}{t + 1}$

**2** (a) $\dfrac{3}{x} - \dfrac{6}{2x + 1}$

  (b) $\dfrac{9}{8(x - 5)} - \dfrac{1}{8(x + 3)}$

  (c) $\dfrac{1}{5(x - 2)} + \dfrac{6}{5(4x - 3)}$

  (d) $\dfrac{1}{2x - 3} - \dfrac{1}{2x + 3}$

  (e) $\dfrac{4}{9(x - 8)} - \dfrac{4}{9(x + 1)}$

  (f) $\dfrac{1}{x - 2} + \dfrac{1}{2(x + 1)}$

**3** $-\dfrac{8}{9x} + \dfrac{1}{3x^2} + \dfrac{8}{9(x - 3)}$

**4** $\dfrac{1}{x + 4} - \dfrac{x}{x^2 + 1}$

**5** $\dfrac{1}{2(x + 3)} + \dfrac{2}{(x - 1)^2} - \dfrac{1}{2(x - 1)}$

**6** $\dfrac{1}{4(x - 1)} - \dfrac{1}{2(x + 1)^2} + \dfrac{3}{4(x + 1)}$

**7** $\dfrac{1}{6(x - 1)} + \dfrac{5 - x}{6(x^2 + 5)}$

**8** $\dfrac{5}{6(x + 2)} - \dfrac{5x - 4}{6(x^2 + 2)}$

**9** (a) $1 + \dfrac{1}{x + 1} - \dfrac{4}{x + 2}$

  (b) $1 + \dfrac{3}{x^2} - \dfrac{3}{x} + \dfrac{2}{x + 1}$

**10** (a) $y = \dfrac{4}{x - 2} - \dfrac{2}{x - 1};$

  $\dfrac{dy}{dx} = -\dfrac{4}{(x - 2)^2} + \dfrac{2}{(x - 1)^2};$

  $\dfrac{d^2y}{dx^2} = \dfrac{8}{(x - 2)^3} - \dfrac{4}{(x - 1)^3}$

  (b) $y = \dfrac{3}{5(x + 3)} + \dfrac{2}{5(x - 2)};$

  $\dfrac{dy}{dx} = -\dfrac{3}{5(x + 3)^2} - \dfrac{2}{5(x - 2)^2};$

  $\dfrac{d^2y}{dx^2} = \dfrac{6}{5(x + 3)^3} + \dfrac{4}{5(x - 2)^3}$

  (c) $y = \dfrac{3}{2(x + 1)} - \dfrac{5}{x + 2} + \dfrac{9}{2(x + 3)};$

  $\dfrac{dy}{dx} = -\dfrac{3}{2(x + 1)^2} + \dfrac{5}{(x + 2)^2} - \dfrac{9}{2(x + 3)^2};$

  $\dfrac{d^2y}{dx^2} = \dfrac{3}{(x + 1)^3} - \dfrac{10}{(x + 2)^3} + \dfrac{9}{(x + 3)^3}$

## Exercise 11e

1  $1 - x + \frac{x^2}{2} - \frac{x^3}{2}, -\frac{1}{2} < x < \frac{1}{2}$

2  $1 - 10x + 75x^2 - 500x^3, -\frac{1}{5} < x < \frac{1}{5}$

3  $1 + \frac{3}{2}x + \frac{3}{2}x^2 + \frac{5}{4}x^3, -2 < x < 2$

4  $1 + \frac{3}{2}x + \frac{3}{8}x^2 - \frac{1}{16}x^3, -1 < x < 1$

5  $\frac{1}{3} - \frac{x}{9} + \frac{x^2}{27} - \frac{x^3}{81}, -3 < x < 3$

6  $1 - \frac{x}{4} + \frac{3x^2}{32} - \frac{5x^3}{128}, -2 < x < 2$

7  $1 + 2x + 3x^2 + 4x^3, -1 < x < 1$

8  $1 - \frac{1}{2}x + \frac{3}{8}x^2 - \frac{5}{16}x^3, -1 < x < 1$

9  $1 + \frac{1}{2}x - \frac{5}{8}x^2 - \frac{3}{16}x^3, -1 < x < 1$

10  $-2 - 3x - 3x^2 - 3x^3, -1 < x < 1$

11  $2 + 2x + \frac{21}{4}x^2 + \frac{27}{2}x^3, -\frac{1}{3} < x < \frac{1}{3}$

12  $\frac{1}{2} - \frac{3}{4}x + \frac{13}{8}x^2 - \frac{51}{16}x^3, -\frac{1}{2} < x < \frac{1}{2}$

13  $1 + x + \frac{1}{2}x^2 + \frac{1}{2}x^3, -1 < x < 1$

14  $1 - \frac{1}{9}x^2, -3 < x < 3$

15  $x + x^2 + 3x^3, -\frac{1}{2} < x < \frac{1}{2}$

16  $x - x^2 + x^3, -1 < x < 1$

17  $1 - 3p^{-1} + 6p^{-2} - 10p^{-3} + 15p^{-4}, |p| < 1$

18  $1 + 2x + 2x^2, -\frac{1}{2} < x < \frac{1}{2}$

19  $1 - 3x + \frac{7}{2}x^2$

21  $1 + 2x + 5x^2$

## Chapter 12

## Exercise 12a

Add a constant to each indefinite integral.

1  $2 \ln \left| \frac{x}{x+1} \right|$

2  $\ln \left| \frac{x-2}{x+2} \right|$

3  $\frac{1}{2} \ln |x^2 - 1|$

4  $\frac{1}{2} \ln \left| \frac{(x+2)^3}{x} \right|$

5  $\ln \frac{(x-3)^2}{|x-2|}$

6  $\frac{1}{2} \ln \frac{|x^2-1|}{x^2}$

7  $\frac{1}{2} \ln |x^2 - 1|$

8  $\ln \left| \frac{x-1}{x+1} \right|$

9  $\ln \frac{(x-3)^6}{(x-2)^4}$

10  $\ln \left| \frac{(x-3)^3}{x-2} \right|$

11  $\frac{1}{2} \ln \frac{5}{3}$

12  $2 \ln 2 - \frac{3}{2} \ln 3$

13  $\ln \frac{1}{6}$

14  $\frac{1}{4} \ln \frac{432}{125}$

## Exercise 12b

1  $\ln (4 + \sin x)$

2  $\frac{1}{3} \ln |3e^x - 1|$

3  $-\frac{1}{2} \ln |(1 - x^2)|$

4  $\ln |\sin x|$

5  $\frac{1}{4} \ln (1 + x^4)$

6  $\ln |x^2 + 3x - 4|$

7  $\ln (\ln x)$

8  $-\ln |1 - x^2|$

9  $-\ln |\cos x|$

10  $\ln |4 + \sec x|$

11  $\frac{1}{2} \ln |x(x - 2)|$

12  $\ln |e^x - x|$

13  $\ln 3$

14  $\ln \sqrt{2}$

## Exercise 12c

1  $x \sin x + \cos x$

2  $\frac{1}{2}(x - 2)e^{2x}$

3  $\frac{1}{16}x^4(4 \ln |3x| - 1)$

4  $-e^{-x}(x + 1)$

5  $3(\sin x - x \cos x)$

6  $e^x (2 - x)$

7  $\frac{1}{2}e^x(\sin x - \cos x)$

8  $(x - 1)e^{x-1}$

9  $e^{2x} (1 - x)$

10  $x(\ln |2x| - 1)$

11  $xe^x$

12  $\frac{1}{72}(8x - 1)(x + 1)^8$

13  $\sin \left( x + \frac{1}{6}\pi \right) - x \cos \left( x + \frac{1}{6}\pi \right)$

14  $\frac{1}{n^2}(\cos nx + nx \sin nx)$

15  $\frac{x^{n+1}}{(n+1)^2}[(n + 1) \ln |x| - 1]$

16  $\frac{3}{4}(2x \sin 2x + \cos 2x)$

17  $(3x - 2) \sin x + 3 \cos x$

18  $\frac{1}{2}(\ln |x|)^2$

19  $e^{\frac{x}{2}}(4x - 8)$

20  $\frac{2}{3}x^{\frac{3}{2}}(\ln |x| - 2)$

21  $e^x (x^2 - 2x + 2)$

22  $-x^2 \cos x + 2x \sin x + 2 \cos x$

23  $\dfrac{e^{4x}}{32}(8x^2 - 4x + 1)$

24  $\dfrac{e^{2x}}{5}(\sin x + 2\cos x)$

## Exercise 12d

1  1

2  $\dfrac{32}{3}\ln 2 - \dfrac{7}{4}$

3  e

4  $-4$

5  $\dfrac{16}{15}$

6  $2\ln x - \dfrac{3}{4}$

7  $2\ln 2 - 1$

8  $e - 2$

9  $\dfrac{1}{4}$

## Exercise 12e

1  (a)  $\dfrac{-1}{\sqrt{1 - x^2}}$

(b)  $\dfrac{2}{\sqrt{1 - 4x^2}}$

(c)  $\dfrac{1}{1 + x^2}$

(d)  $x\cos^{-1} x - \sqrt{1 - x^2}$

(e)  $x\sin^{-1} 2x + \dfrac{1}{2}\sqrt{1 - 4x^2}$

(f)  $x\tan^{-1} x - \dfrac{1}{2}\ln(1 + x^2)$

## Exercise 12f

1  $\dfrac{1}{10}(x^2 - 3)^5$

2  $-\dfrac{1}{3}(1 - x^2)^{\frac{3}{2}}$

3  $\dfrac{1}{6}(\sin 2x + 3)^3$

4  $-\dfrac{1}{6}(1 - x^3)^2$

5  $\dfrac{2}{3}(1 + e^x)^{\frac{3}{2}}$

6  $\dfrac{1}{5}\sin^5 x$

7  $\dfrac{1}{4}\tan^4 x$

8  $\dfrac{1}{3(n+1)}(1 + x^{n+1})^3$

9  $-\dfrac{1}{3}\cot^3 x$

10  $\dfrac{4}{9}\left(1 + x^{\frac{3}{2}}\right)^{\frac{3}{2}}$

11  $\dfrac{1}{21}(x + 3)^6(3x + 2)$

12  $-\dfrac{2}{3}(x + 6)\sqrt{(3 - x)}$

13  $\dfrac{2}{15}(3x - 2)(x + 1)^{\frac{3}{2}}$

14  $\dfrac{1 - 5x}{10(x - 3)^5}$

15  $\dfrac{4}{135}(9x + 8)(3x - 4)^{\frac{3}{2}}$

16  $-\dfrac{1}{36}(8x + 1)(1 - x)^8$

17  $\dfrac{5 + 4x}{12(4 - x)^4}$

18  $-2\sqrt{7 + \cos x}$

19  $\dfrac{1}{12}(x^4 + 4)^3$

20  $-\dfrac{1}{4}(1 - e^x)^4$

21  $\dfrac{2}{3}(1 - \cos\theta)^{\frac{3}{2}}$

22  $\dfrac{1}{3}(x^2 + 2x + 3)^{\frac{3}{2}}$

23  $\dfrac{1}{2}e^{(x^2 + 1)}$

24  $\dfrac{1}{2}(1 + \tan x)^2$

25  $\dfrac{1}{2}\ln\left|\dfrac{\cos x - 1}{\cos x + 1}\right|$

## Exercise 12g

1  $\dfrac{1}{2}(e - 1)$

2  $\dfrac{1}{5}$

3  $\dfrac{1}{2}(\ln 2)^2$

4  $\dfrac{7^5}{15}$

5  $e - 1$

6  9

7  $\dfrac{1}{2}(e^3 - 1)$

8  $\dfrac{13}{24}$

9  $\dfrac{1}{3}(\ln 3)^3$

10  $\dfrac{7}{3}$

## Mixed exercise 12

1  $\dfrac{1}{10}(1 + x^2)^5$

2  $-\dfrac{1}{9}e^{-3x}(3x + 1)$

3  $\dfrac{1}{4}e^{2x}(2x^2 - 2x + 1)$

4  $x + \ln|x + 2|$

5  $\dfrac{-1}{3(x^3 + 1)}$

6  $\ln\left|\dfrac{x - 4}{x - 1}\right|$

7  $\ln\left|\dfrac{x}{\sqrt{2x + 1}}\right|$

8  $\ln\sqrt{x^2 + 1} - \tan^{-1} x$

9  $-2\sqrt{\cos x}$

10  $\sqrt{2}$

11  $\dfrac{256}{15}$

12  $\ln\sqrt{2}$

13  $\dfrac{7}{3}$

14  $\ln\dfrac{9}{4}$

15  $\ln\dfrac{5}{2}$

16  $\ln\dfrac{3}{8}$

# Chapter 13

## Exercise 13a

1 $y^2 = A - 2\cos x$

2 $\dfrac{1}{y} - \dfrac{1}{x} = A$

3 $2y^3 = 3(x^2 + 4y + A)$

4 $x = A\sec y$

5 $(A - x)y = 1$

6 $y = \ln\dfrac{A}{\sqrt{1 - x^2}}$

7 $y = A(x - 3)$

8 $x + A = 4\ln|\sin y|$

9 $u^2 = v^2 + 4v + A$

10 $y^2 + 2(x + 1)e^{-x} = A$

11 $u + 2 = A(v + 1)$

12 $y^2 = Ax(x + 2)$

13 $4v^3 = 3(2 + t)^4 + A$

14 $1 + y^2 = Ax^2$

15 $y^2 = A - \operatorname{cosec}^2 x$

16 $v^2 + A = 2u - 2\ln|u|$

17 $e^{-x} = e^{1-y} + A$

## Exercise 13b

1 $y^3 = x^3 + 3x - 13$

2 $e^t(5 - 2\sqrt{s}) = 1$

3 $3(y^2 - 1) = 8(x^2 - 1)$

4 $y = x^2 - x$

5 $y = e^x - 2$

6 $y = 5 - \dfrac{3}{x}$

7 $\left(0, \sqrt{3 + e^{-3}}\right), \left(0, -\sqrt{3 + e^{-3}}\right)$

8 $(y + 1)^2(x + 1) = 2(x - 1)$

9 $2y = x^2 + 6x$

10 $4y^2 = (y + 1)^2(x^2 + 1)$

11 $x^3y = y - 1$

12 $y = \tan\left\{\tfrac{1}{2}(x^2 - 4)\right\}$

## Exercise 13c

1 $s\dfrac{ds}{dt} = k$

2 $\dfrac{dh}{dt} = k\ln|H - h|$

3 (a) $\dfrac{dn}{dt} = k_1 n$    (b) $\dfrac{dn}{dt} = \dfrac{k_2}{n}$    (c) $\dfrac{dn}{dt} = -k_3$

## Exercise 13d

1 $t = \dfrac{16T}{3}$

2 $\dfrac{dy}{dx} = k\sqrt{x};\ y = 0.4x^{\frac{3}{2}} + 1.6;\ 1.2$

3 (a) $\dfrac{dn}{dt} = kn$

(b) $t = \dfrac{10\ln 2}{\ln 1.5} = 1.71$ (hours)

4 (a) $-\dfrac{dm}{dt} = km;\ m = 50e^{-kt}$
where $k = 0.002554\ldots$

(b) $26.8\,\text{g}$ (3 s.f.)

5 (c) $k = 0.0357$ (3 s.f.)

(d) $44°$ (nearest degree)

6 (a) $\dfrac{dN}{dt} = kN$    (b) 33

# Chapter 14

## Exercise 14a

1 (a) $2\mathbf{i} - \mathbf{j} - 5\mathbf{k}$    (b) $3\mathbf{j} - 5\mathbf{k}$
(c) $-2\mathbf{i} + 4\mathbf{j} + \mathbf{k}$    (d) $\mathbf{i} + \mathbf{k}$

2 (a) $\mathbf{r} = \mathbf{i} - 3\mathbf{j} + 2\mathbf{k} + t(5\mathbf{i} + 4\mathbf{j} - \mathbf{k})$
(b) $\mathbf{r} = 2\mathbf{i} + \mathbf{j} + t(3\mathbf{j} - \mathbf{k})$
(c) $\mathbf{r} = t(\mathbf{i} - \mathbf{j} - \mathbf{k})$

3 (a) no    (b) yes    (c) yes    (d) yes

4 (a) $\mathbf{r} = 4\mathbf{i} + 5\mathbf{j} + 10\mathbf{k} + s(\mathbf{i} + \mathbf{j} + 3\mathbf{k})$
$\mathbf{r} = 2\mathbf{i} + 3\mathbf{j} + 4\mathbf{k} + s(\mathbf{i} + \mathbf{j} + 5\mathbf{k})$
$\mathbf{r} = 4\mathbf{i} + 5\mathbf{j} + 10\mathbf{k} + s(\mathbf{i} + \mathbf{j} + 5\mathbf{k})$

(b) 1.09

5 (a) $\mathbf{r} = \mathbf{i} + 7\mathbf{j} + 8\mathbf{k} + s(3\mathbf{i} + \mathbf{j} - 4\mathbf{k})$
$(7, 9, 0), \left(0, \tfrac{20}{3}, \tfrac{28}{3}\right), (-20, 0, 36)$
(b) $\mathbf{r} = \mathbf{i} + \mathbf{j} + 7\mathbf{k} + s(2\mathbf{i} + 3\mathbf{j} - 6\mathbf{k})$
$\left(\tfrac{10}{3}, \tfrac{9}{2}, 0\right), \left(0, -\tfrac{1}{2}, 10\right), \left(\tfrac{1}{3}, 0, 9\right)$

6 1.67

## Exercise 14b

1 (a) parallel    (b) intersect at $\mathbf{i} + 2\mathbf{j}$; $27.9°$
(c) skew

2 $-3, -\mathbf{i} + 3\mathbf{j} + 4\mathbf{k}, 64.6°$

## Exercise 14c

1 (a) $x + y - z = 2$    (b) $2x + 3y - 4z = 1$

2 (a) $\mathbf{r}.(3\mathbf{i} - 2\mathbf{j} + \mathbf{k}) = 5$    (b) $\mathbf{r}.(5\mathbf{i} - 3\mathbf{j} - 4\mathbf{k}) = 7$

3 1. (a) $\mathbf{i} + \mathbf{j} - \mathbf{k}$    (b) $2\mathbf{i} + 3\mathbf{j} - 4\mathbf{k}$
2. (a) $3\mathbf{i} - 2\mathbf{j} + \mathbf{k}$    (b) $5\mathbf{i} - 3\mathbf{j} - 4\mathbf{k}$

4 1. (a) $\dfrac{2}{\sqrt{3}}$    (b) $\dfrac{1}{\sqrt{29}}$
2. (a) $\dfrac{5}{\sqrt{14}}$    (b) $\dfrac{7}{\sqrt{50}}$

5 The normals are parallel, 2 units

6 $\mathbf{r} = s(\mathbf{i} - 2\mathbf{j} + \mathbf{k})$

7 $\mathbf{r} = 2\mathbf{i} + \mathbf{j} + \mathbf{k} + t(\mathbf{i} + 2\mathbf{j} - 3\mathbf{k})$

8 $\mathbf{r}.(7\mathbf{i} + 2\mathbf{j} - 3\mathbf{k}) = 3, \dfrac{3}{\sqrt{62}}$

9 $\mathbf{r}.(\mathbf{i} - \mathbf{j} - 3\mathbf{k}) = -4$

10 $\mathbf{r}.(-\mathbf{i} + \mathbf{k}) = 3$

11  $\left(\frac{5}{2}, -\frac{1}{2}, -\frac{1}{2}\right)$

12  (a)  $6x + 4y + 3z = 12$

　　(b)  $\dfrac{12}{\sqrt{61}}$　　　　　(c)  30.8°

13  (a)  $\mathbf{r} = 2\mathbf{i} + \mathbf{j} - 4\mathbf{k} + t(\mathbf{i} - \mathbf{j} + 3\mathbf{k})$

　　(b)  $\frac{1}{10}(33\mathbf{i} - 3\mathbf{j} - \mathbf{k})$

　　(c)  33.4°

14  (a)  $\mathbf{r} = 2\mathbf{i} + \mathbf{j} - 2\mathbf{k} + t(3\mathbf{j} + 3\mathbf{k})$

　　(b)  90°

15  (a)  $\mathbf{r} = 5\mathbf{j} + 3\mathbf{k} + t(\mathbf{i} - 2\mathbf{j} - \mathbf{k})$

　　(b)  15.2°

16  $2x - y - 2z = 9$

## Chapter 15

### Exercise 15a

1  $-i, i, i, -i, 1, i$

2  (a)  $10 + 4i$　　　　(b)  $7 + 2i$
　　(c)  $6 - 2i$　　　　(d)  $(a + c) + (d - b)i$

3  (a)  $-4 + 6i$　　　　(b)  $1 - 4i$
　　(c)  $-2 + 16i$　　　(d)  $(a - c) - (b + d)i$

4  (a)  $10 - 5i$　　　　(b)  $39 + 23i$
　　(c)  $11 - 7i$　　　　(d)  $25$
　　(e)  $3 - 4i$　　　　(f)  $-2 + 2i$
　　(g)  $-4 + 3i$　　　　(h)  $x^2 + y^2$
　　(i)  $-3 + i$　　　　(j)  $(a^2 - b^2) + 2abi$

5  (a)  $1 + i$　　　　(b)  $\frac{9}{25} + \frac{13}{25}i$
　　(c)  $\frac{4}{17} + \frac{16}{17}i$　　(d)  $i$
　　(e)  $-i$　　　　(f)  $\dfrac{x^2 - y^2}{x^2 + y^2} + \dfrac{2xy}{x^2 + y^2}i$
　　(g)  $1 - 3i$　　　　(h)  $-3 - 2i$

6  (a)  $x = 9, y = -7$
　　(b)  $x = -\frac{3}{2}, y = \frac{7}{2}$
　　(c)  $x = \frac{7}{2}, y = \frac{1}{2}$
　　(d)  $x = 2, y = 0$
　　(e)  $x = 13, y = 0$
　　(f)  $x = 15, y = 8$
　　(g)  $x = 11, y = 3$
　　(h)  $x = 2, y = 1$ or $x = -2, y = -1$

7  (a)  $7, -1$　　　　(b)  $-2, 2$
　　(c)  $\frac{10}{17}, \frac{11}{17}$　　　(d)  $\frac{9}{5}, -\frac{4}{5}$
　　(e)  $0, \dfrac{-2y}{x^2 + y^2}$　　(f)  $-1, 0$
　　(g)  $-\frac{1}{2}, \frac{1}{2}\sqrt{3}$

### Exercise 15b

1  $2 - i$ and $-2 + i$

2  $1 + i$ and $-(1 + i)$

3  $2 + i$ and $-2 - i$

4  $\sqrt{2} + i\sqrt{3}$ and $-(\sqrt{2} + i\sqrt{3})$

5  $2.20 - 0.910i$ and $-2.20 + 0.910i$

### Exercise 15c

1  (a)  $-\frac{1}{2} \pm \frac{1}{2}\sqrt{3}i$　　(b)  $-\frac{7}{4} \pm \frac{1}{4}\sqrt{41}$
　　(c)  $\pm 3i$　　　　(d)  $-\frac{1}{2} \pm \frac{1}{2}\sqrt{11}i$
　　(e)  $\pm 1, \pm i$

2  (a)  $x^2 + 1 = 0$
　　(b)  $x^2 - 4x + 5 = 0$
　　(c)  $x^2 - 2x + 10 = 0$
　　(d)  $x^3 - 4x^2 + 6x - 4 = 0$

3  $a = 9$

### Exercise 15d

1  (a)  $\sqrt{13}, -0.59$　　(b)  $\sqrt{17}, -0.24$
　　(c)  $5, -2.21$　　　（d)  $13, 1.18$
　　(e)  $\sqrt{2}, -\frac{1}{4}\pi$　　(f)  $\sqrt{2}, \frac{3}{4}\pi$
　　(g)  $4, 0$　　　　(h)  $2, -\frac{1}{2}\pi$
　　(i)  $\sqrt{a^2 + b^2}, \tan^{-1}\dfrac{b}{a}$
　　(j)  $\sqrt{2}, \frac{1}{4}\pi$　　(k)  $\sqrt{2}, \frac{3}{4}\pi$
　　(l)  $\sqrt{2}, -\frac{3}{4}\pi$　　(m)  $\sqrt{2}, -\frac{1}{4}\pi$
　　(n)  $\sqrt{170}, 0.57$　　(o)  $2, \frac{1}{3}\pi$
　　(p)  $1, \frac{3}{4}\pi$　　　(q)  $3, -\frac{5}{6}\pi$

2  (a)  $4 + 3i$　　　　(b)  $5 + 3i$
　　(c)  $7 - 2i$　　　　(d)  $-1 - i$
　　(e)  $-5 - 4i$　　　（f)  $-2 + 6i$
　　(i)  $-2 + 5i$　　　（j)  $-1$

3  (a)  $\sqrt{2}\left(\cos\dfrac{\pi}{4} + i\sin\dfrac{\pi}{4}\right)$
　　(b)  $2\left\{\cos\left(-\dfrac{\pi}{6}\right) + i\sin\left(-\dfrac{\pi}{6}\right)\right\}$
　　(c)  $5\{\cos(-2.214^c) + i\sin(-2.214^c)\}$
　　(d)  $13(\cos 1.966^c + i\sin 1.966^c)$
　　(e)  $\sqrt{5}\{\cos(-0.464^c) + i\sin(-0.464^c)\}$
　　(f)  $6(\cos 0 + i\sin 0)$
　　(g)  $3(\cos\pi + i\sin\pi)$
　　(h)  $4\left(\cos\dfrac{\pi}{2} + i\sin\dfrac{\pi}{2}\right)$
　　(i)  $2\sqrt{3}\left\{\cos\left(-\dfrac{5\pi}{6}\right) + i\sin\left(-\dfrac{5\pi}{6}\right)\right\}$
　　(j)  $25(\cos 0.284^c + i\sin 0.284^c)$

4  (a)  $\sqrt{3} + i$
　　(b)  $\frac{3}{2}\sqrt{2} - \frac{3}{2}\sqrt{2}i$
　　(c)  $-\frac{1}{2} + \frac{1}{2}\sqrt{3}i$
　　(d)  $-\frac{1}{2}\sqrt{2} - \frac{1}{2}\sqrt{2}i$
　　(e)  $3$
　　(f)  $-2$
　　(g)  $2\sqrt{3} - 2i$
　　(h)  $-1$
　　(i)  $-3i$
　　(j)  $-\dfrac{1}{2} - \dfrac{i\sqrt{3}}{2}$

5　(a)　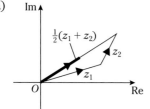

　　(b)　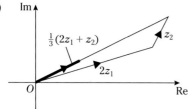

## Exercise 15e

1　(a)　$2\sqrt{2}, \frac{1}{4}\pi$

　　(b)　$2\sqrt{6}, -\frac{5}{12}\pi$

　　(c)　$1, 2\tan^{-1}\sqrt{\frac{3}{2}}$

2　$\pm 2\left\{\cos\left(-\frac{\pi}{12}\right) + i\sin\left(-\frac{\pi}{12}\right)\right\}$

3　$9, \frac{3\pi}{4}$

4　$z_1 = e^{\frac{i\pi}{3}}, z_2 = 3e^{\frac{i\pi}{6}}$,

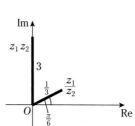

## Exercise 15f

1　Circle, centre $O$, radius 1

2　Circle, centre $(1, 0)$, radius 3

3　Circle, centre $(0, 2)$, radius 3

4　Circle, centre $(0, -2)$, radius 2

5　Circle, centre $(1, -1)$, radius 4

6　Straight line through $O$ at $\frac{\pi}{3}$ to $Ox$

7　The perpendicular bisector of the line joining $(2, 3)$ to $(-4, 5)$

8　The perpendicular bisector of the line joining $O$ to $(0, -4)$

9　(a)　Inside the circle centre $(1, 0)$, radius 4

　　(b)　Outside the circle centre $(0, -3)$, radius 2

　　(c)　Inside the circle centre $(-1, 1)$, radius 1

　　(d)　Between the lines through $O$ inclined at $\frac{\pi}{3}$ and $\frac{2\pi}{3}$

10　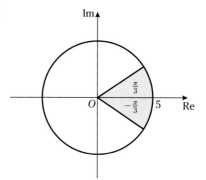

11　(a)　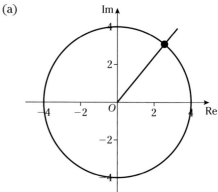

　　(b)　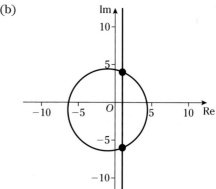

12　$\frac{\sqrt{2}}{2} - \frac{\sqrt{2}}{2}i$

13　(a)　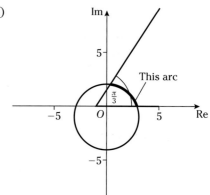

(b)

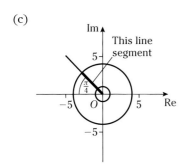

(c) $\dfrac{(\sqrt{2}+1)}{(\sqrt{2}-1)}$

10 (a) $\frac{1}{2}, -\frac{1}{2}\pi$

(b) $3 - 2i, -3 + 2i$;

Diagram showing $\pm(2 + 3i)\sqrt{3}$

## Summary exercise 3

1 (i) $-\frac{1}{2} + \dfrac{\sqrt{3}i}{2}, -\sqrt{3} - i, \sqrt{3} - i$

2 $\dfrac{1}{2x + 1} + \dfrac{4}{x - 2} + \dfrac{8}{(x - 2)^2}$

3 (ii) $x = 100 - 95e^{-0.2t}$

(iii) $x \to 100$

4 (i) $40.4°$

(ii) $\mathbf{r} = 3\mathbf{j} + 2\mathbf{k} + t(6\mathbf{i} - 10\mathbf{j} - 7\mathbf{k})$

5 $y = \sqrt[3]{2e^{3x} - 1}$

6 (i) $\frac{1}{2} + \dfrac{\sqrt{3}i}{2}, \frac{1}{2} - \dfrac{\sqrt{3}i}{2}$ (ii) $1, \dfrac{\pi}{3}$ and $1, -\dfrac{\pi}{3}$

7 (i) $\dfrac{-1}{x - 1} + \dfrac{4}{x - 2} - \dfrac{2}{x + 1}$

8 (i) $(1, 0)$ (ii) $x = e^{0.5}, y = \frac{1}{2}e^{-1}$

(iii) $\dfrac{e - 2}{e}$

(c)

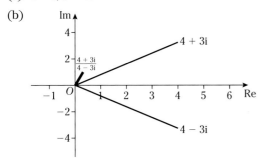

9 $\frac{1}{4} - \frac{3}{4}x + \frac{27}{16}x^2$

10 $y = x$

11 $0.802$

12 (i) $2, \dfrac{\pi}{3}$

13 (ii) $h = 9 - \left(4 - \frac{1}{15}t\right)^{\frac{3}{2}}$

(iii) $9, 60$ years (iv) $19.1$ years

14 (ii) $5\mathbf{i} + 3\mathbf{j} + 4\mathbf{k}$

15 (ii) $0.76$

16 (ii) $24.7°, 95.3°$

17 (i) $-3$ (ii) $-\dfrac{10}{3}x^3$

18 (i) $0.98$

(ii) Less than $E$

19 (i) $1 - \sqrt{3}i, -1 - \sqrt{3}i$

(iii) $2, -\dfrac{\pi}{3}$ and $2, -\dfrac{2\pi}{3}$

20 (i) $\dfrac{1}{x} + \dfrac{10}{x^2} + \dfrac{1}{10 - x}$

(ii) $t = \ln\left(\dfrac{9x}{10 - x}\right) - \dfrac{10}{x} + 10$

21 (i) $b = -2, c = 3$

22 (ii) $y + 4x + 1 = 0$

24 (i) $a = -2$ (ii) $3$

25 (i) $A = 1, B = 2, C = 1, D = -3$

26 (ii) $r = \dfrac{5}{(1 - 0.4t)}$

(iii) $0 \leqslant t < 2.5$

27 $0.365, 1.206$

28 (i) $2.78$

(ii) $x = \dfrac{3}{4}x + \dfrac{15}{x^3}, \sqrt[4]{60}$

## Mixed exercise 15

1 $\frac{1}{5}(3 - 4i), \frac{1}{2}(-3 + i)$

Diagram showing the points $-3i$ and $6 - 5i$

2 $2 - i, 3 - 4i$, a circle centre $(2 - i)$, radius $\sqrt{5}$

3 (a) $\dfrac{104}{25} - \dfrac{72}{25}i$

(b) $\pm\sqrt{2}(1 + i)$

4 $a = \dfrac{63}{25}, b = \dfrac{16}{25}, r = \dfrac{13}{5}, \cos\theta = \dfrac{63}{65}, \sin\theta = \dfrac{16}{65}$

5 $4, -5; 5$

6 $2, \frac{1}{6}\pi; \frac{1}{2}\left(1 + i\sqrt{3}\right)$

7 (a) $3 - i, 3 + 3i$

(b)

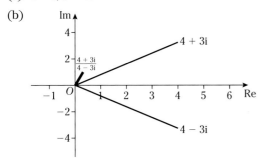

8 (a) $6$

(b) $\dfrac{x(x^2 + y^2 + 1)}{x^2 + y^2}, \dfrac{y(x^2 + y^2 - 1)}{x^2 + y^2}$;

The points $(x + iy)$ where $y(x^2 + y^2 - 1) = 0$

9 (a) $\pm(3 + 2i)$

(b) (i) $\sqrt{2}, -\frac{1}{4}\pi$

(ii) $5, 0.643^c$

(iii) $5\sqrt{2}, -0.142^c; 12$

**29** (i) $\dfrac{1}{x+1} + \dfrac{2}{(x+1)^2} - \dfrac{3}{3x+2}$

(ii) $\dfrac{3}{2} - \dfrac{11}{4}x + \dfrac{29}{8}x^2$

**30** (iv) 2.31

**31** (i) $57.7°$

(ii) $\mathbf{r} = 2\mathbf{i} - \mathbf{k} + \lambda(4\mathbf{i} - 7\mathbf{j} + 5\mathbf{k})$

**32** $1 - \dfrac{3}{2}x^2$

**33** (i) 3

(ii) $x < -\dfrac{1}{2}$

## P2 Sample paper 1

**1** $-1.35$

**2** $-6 < x < \dfrac{2}{3}$

**4** $y = \dfrac{1}{\sqrt{3}}x$

**5** (i) $\dfrac{1}{3}\pi,\ \pi,\ \dfrac{5}{3}\pi$

**6** (iv) 1.75, 1.78885, 1.76168, 1.78055, 1.76739, 1.77654; 1.77

**7** (ii) 3.98, underestimate

(iii) $1 + \ln x$

(iv) $5\ln 5 - 4$

## P2 Sample paper 2

**1** 2, 6

**2** (ii) $2\pi\ \text{unit}^3$

**3** $-9;\ -8,\ -\dfrac{1}{2},\ 1$

**4** (i) $5\cos(x - 36.87°)$  (ii) $103.3°,\ 330.5°$

**5** (iii) 1.1, 1.1545, 1.1468, 1.1479, 1.1477, 1.1478; 1.148

**6** (i) $\dfrac{\sin x}{2\sqrt{2 - \cos x}}$

## P3 Sample paper 1

**1** $-7 < x < 1$

**2** $(-3, 4),\ (2, -1)$

**4** (iii) 1, 1.0472, 1.0742, 1.0896, 1.0982, 1.1030, 1.1057, 1.1072, 1.1081, 1.1085; 1.11

**5** (ii) $\dfrac{1}{8}\sqrt{3} + \dfrac{1}{12}\pi$

**6** (i) (b) $\dfrac{1}{4}\pi,\ \dfrac{5}{6}\pi$

**7** (i) $\mathbf{r} = -2\mathbf{i} + 2\mathbf{j} + 4\mathbf{k} + t(2\mathbf{i} + \mathbf{j} - 3\mathbf{k})$

(ii) $3\mathbf{j} + \mathbf{k}$

(iii) $4x - 5y + z = -14$

**8** (i) $\dfrac{1}{2+x} - \dfrac{x}{x^2+1}$

(ii) $\dfrac{1}{2} - \dfrac{5}{4}x + \dfrac{1}{8}x^2 + \dfrac{15}{16}x^2$

**9** (i) $\left(\sqrt{e},\ \dfrac{1}{2e}\right)$    (ii) $\dfrac{4}{5} - \dfrac{1}{5}\ln 5\ \text{unit}^2$

**10** (ii) 8.55 days

## P3 Sample paper 2

**1** (ii) $(9, 2)$

**2** $(9, -6)$

**3** (ii) 4, 4.375, 4.358928, 4.35889, 4.35889; 4.3589

**4** $66.4°,\ 120°,\ 240°,\ 293.6°$

**5** (ii) $y = x - 3$

**6** $T = 10 + 90\exp\left(-\dfrac{1}{10}\ln 3t\right)$

**7** (i) $(2, -3, 4)$    (ii) $x - 5y - 3z = 5$

**8** (i) $(1 + i),\ -(1 + i)$    (ii) $\sqrt{2}e^{i\frac{\pi}{4}}\ \sqrt{2}e^{i\frac{5\pi}{4}}$

**9** (i) $1 + \dfrac{1}{2}x^2 + \dfrac{3}{8}x^4 + \dfrac{5}{16}x^6$

(ii) 1.73

**10** (i) $\dfrac{2}{x+3} - \dfrac{2x}{x^2+4}$

# Index